Statistical Computing
in Nuclear Imaging

Series in Medical Physics and Biomedical Engineering

Series Editors: John G Webster, E Russell Ritenour, Slavik Tabakov, and Kwan-Hoong Ng

Other recent books in the series:

Radiosensitizers and Radiochemotherapy in the Treatment of Cancer
Shirley Lehnert

The Physiological Measurement Handbook
John G Webster (Ed)

Diagnostic Endoscopy
Haishan Zeng (Ed)

Medical Equipment Management
Keith Willson, Keith Ison, and Slavik Tabakov

Targeted Muscle Reinnervation: A Neural Interface for Artificial Limbs
Todd A Kuiken; Aimee E Schultz Feuser; Ann K Barlow (Eds)

Quantifying Morphology and Physiology of the Human Body Using MRI
L Tugan Muftuler (Ed)

Monte Carlo Calculations in Nuclear Medicine, Second Edition: Applications in Diagnostic Imaging
Michael Ljungberg, Sven-Erik Strand, and Michael A King (Eds)

Vibrational Spectroscopy for Tissue Analysis
Ihtesham ur Rehman, Zanyar Movasaghi, and Shazza Rehman

Webb's Physics of Medical Imaging, Second Edition
M A Flower (Ed)

Correction Techniques in Emission Tomography
Mohammad Dawood, Xiaoyi Jiang, and Klaus Schäfers (Eds)

Physiology, Biophysics, and Biomedical Engineering
Andrew Wood (Ed)

Proton Therapy Physics
Harald Paganetti (Ed)

Practical Biomedical Signal Analysis Using MATLAB®
K J Blinowska and J Żygierewicz (Ed)

Series in Medical Physics and Biomedical Engineering

Statistical Computing in Nuclear Imaging

Arkadiusz Sitek

Associate Physicist
Massachusetts General Hospital, Department of Radiology
Boston, USA

CRC Press
Taylor & Francis Group
Boca Raton London New York

CRC Press is an imprint of the
Taylor & Francis Group, an **informa** business

CRC Press
Taylor & Francis Group
6000 Broken Sound Parkway NW, Suite 300
Boca Raton, FL 33487-2742

First issued in paperback 2020

ISBN-13: 978-1-4398-4934-7 (hbk)
ISBN-13: 978-0-367-78363-1 (pbk)

Visit the Taylor & Francis Web site at
http://www.taylorandfrancis.com

and the CRC Press Web site at
http://www.crcpress.com

Dedication

To Sylwia and Pascal Jan

Contents

List of Figures xi

List of Tables xvi

About the Series xviii

Preface xxi

About the Author xxiii

Chapter 1   Basic statistical concepts 1

    1.1   Introduction 1
    1.2   Before- and after-the-experiment concepts....................2
    1.3   Definition of probability...6
        1.3.1   Countable and uncountable quantities 8
    1.4   Joint and conditional probabilities.............................10
    1.5   Statistical model 13
    1.6   Likelihood 17
    1.7   Pre-posterior and posterior 19
        1.7.1   Reduction of pre-posterior to posterior 19
        1.7.2   Posterior through Bayes theorem 19
        1.7.3   Prior selection...20
        1.7.4   Examples 22
        1.7.5   Designs of experiments 23
    1.8   Extension to multi–dimensions 25
        1.8.1   Chain rule and marginalization 26
        1.8.2   Nuisance quantities..27
    1.9   Unconditional and conditional independence...............29
    1.10 Summary..34

Chapter 2   Elements of decision theory 37

    2.1   Introduction 37
    2.2   Loss function and expected loss 39
    2.3   After-the-experiment decision making 42
        2.3.1   Point estimation 43
        2.3.2   Interval estimation...48

2.3.3   Multiple-alternative decisions 50
2.3.4   Binary hypothesis testing/detection 52
2.4   Before-the-experiment decision making 56
2.4.1   Bayes risk 58
2.4.2   Other methods 62
2.5   Robustness of the analysis 64

Chapter 3   Counting statistics 67

3.1   Introduction to statistical models 67
3.2   Fundamental statistical law 69
3.3   General models of photon-limited data 71
3.3.1   Binomial statistics of nuclear decay 71
3.3.2   Multinomial statistics of detection 72
3.3.3   Statistics of complete data 77
3.3.4   Poisson-multinomial distribution of nuclear
           data 84
3.4   Poisson approximation 88
3.4.1   Poisson statistics of nuclear decay 88
3.4.2   Poisson approximation of nuclear data 90
3.5   Normal distribution approximation 93
3.5.1   Approximation of binomial law 94
3.5.2   Central limit theorem 95

Chapter 4   Monte Carlo methods in posterior analysis 99

4.1   Monte Carlo approximations of distributions 99
4.1.1   Continuous distributions 99
4.1.2   Discrete distributions 104
4.2   Monte Carlo integrations 107
4.3   Monte Carlo summations 110
4.4   Markov chains 111
4.4.1   Markov processes 113
4.4.2   Detailed balance 114
4.4.3   Design of Markov chain 116
4.4.4   Metropolis–Hastings sampler 118
4.4.5   Equilibrium 120
4.4.6   Resampling methods (bootstrap) 126

Chapter 5   Basics of nuclear imaging 129

5.1   Nuclear radiation 130
5.1.1   Basics of nuclear physics 130
5.1.1.1   Atoms and chemical reactions 130
5.1.1.2   Nucleus and nuclear reactions 131
5.1.1.3   Types of nuclear decay 133

5.1.2   Interaction of radiation with matter.............. 136
        5.1.2.1   Inelastic scattering........................... 137
        5.1.2.2   Photoelectric effect 138
        5.1.2.3   Photon attenuation.......................... 138
5.2   Radiation detection in nuclear imaging 141
      5.2.1   Semiconductor detectors............................... 142
      5.2.2   Scintillation detectors..................................... 143
              5.2.2.1   Photomultiplier tubes...................... 144
              5.2.2.2   Solid-state photomultipliers............. 145
5.3   Nuclear imaging 147
      5.3.1   Photon-limited data 150
      5.3.2   Region of response (ROR)............................. 152
      5.3.3   Imaging with gamma camera 153
              5.3.3.1   Gamma camera............................... 153
              5.3.3.2   SPECT 157
      5.3.4   Positron emission tomography (PET) 159
              5.3.4.1   PET nuclear imaging scanner.......... 159
              5.3.4.2   Coincidence detection 161
              5.3.4.3   ROR for PET and TOF-PET.......... 162
              5.3.4.4   Quantitation of PET 165
      5.3.5   Compton imaging 166
5.4   Dynamic imaging and kinetic modeling..................... 168
      5.4.1   Compartmental model.................................... 169
      5.4.2   Dynamic measurements................................. 171
5.5   Applications of nuclear imaging................................. 173
      5.5.1   Clinical applications 173
      5.5.2   Other applications 174

Chapter 6   Statistical computing 179

6.1   Computing using Poisson-multinomial distribution
      (PMD).. 179
      6.1.1   Sampling the posterior 180
      6.1.2   Computationally efficient priors 182
      6.1.3   Generation of Markov chain 186
      6.1.4   Metropolis–Hastings algorithm 187
      6.1.5   Origin ensemble algorithms 190
6.2   Examples of statistical computing 193
      6.2.1   Simple tomographic system (STS) 194
      6.2.2   Image reconstruction 195
      6.2.3   Bayes factors.. 198
      6.2.4   Evaluation of data quality 200
      6.2.5   Detection—Bayesian decision making 203
      6.2.6   Bayes risk 205

Appendix A Probability distributions 209

    A.1 Univariate distributions 209
        A.1.1 Binomial distribution...................................... 209
        A.1.2 Gamma distribution 209
        A.1.3 Negative binomial distribution 209
        A.1.4 Poisson-binomial distribution 210
        A.1.5 Poisson distribution.. 210
        A.1.6 Uniform distribution....................................... 210
        A.1.7 Univariate normal distribution 210
    A.2 Multivariate distributions 211
        A.2.1 Multinomial distribution 211
        A.2.2 Multivariate normal distribution 211
        A.2.3 Poisson-multinomial distribution.................... 211

Appendix B Elements of set theory 213

Appendix C Multinomial distribution of single-voxel imaging............... 217

Appendix D Derivations of sampling distribution ratios........................ 221

Appendix E Equation (6.11) 223

Appendix F C++ OE code for STS 225

References 231

Index... **239**

List of Figures

1.1   Before-experiment and after-experiment concepts..................................5
1.2   Observable, unobservable, and known quantities (OQs, UQs, KQs)
      are identified. UQs are uncertain at the AE stage. Both OQs and
      UQs are QoIs...6
1.3   (A) 2D joint distribution defined for $\{f, g\}$. (B) Values of the joint
      distribution for QoI $g = 0.3$. (C) Conditional distribution of f
      conditioned on $g = 0.3$. The shapes of both distributions in (B)
      and (C) are identical and they differ only by a scaling factor.............12
1.4   Three distributions extracted from the joint distribution $p(f, g) =$
      $144(f - 0.5)^2(g - 0.5)^2$ for different values of g equal to (A) 0.3,
      (B) 0.8, and (C) 0.0. After normalization the same conditional
      distribution is obtained (D)..14
1.5   The likelihood function for the model described in Example 1.10
      corresponding to $g = 1$ (left) and $g = 5$ (right). The values in the
      figure correspond to rows 1 and 5 in Table 1.1. 18
1.6   Distributions introduced in Chapter 1. It is assumed that OQ is g
      and UQ is f..25
1.7   (A) The wire-cube represents the joint probability of three QoIs
      f, $\tilde{f}$, and g. There are only two possible true values for each QoI
      and each corner of the cube correspond to different values of f, $\tilde{f}$,
      and g, which are shown in (F). The values of each QoI are either
      0 or 1. (B), (C), (D) show some conditionals and marginalizations
      described in detail in Example 1.16.......................................30
1.8   Comparison of $p(f, g)$ and $p(f)p(g)$ for Example 1.17. Lines con-
      necting the points were added for clarity. 33

2.1   (A) Simplified model of the traditional two-step decision-making
      process in interpretation of medical nuclear imaging data. (B) An
      alternative approach in which the formation of images is not used
      and decisions are made based directly on the data. 38
2.2   The urn with three balls. The experiment consists of random draw-
      ing of one of the balls. 40
2.3   Two definitions of the loss function shown in (A) and (B). The
      "true" corresponds to the result of the draw, and $\delta = \text{Yes}$ corre-
      sponds to the decision that the ball with number 1 was drawn. 40
2.4   The one-dimensional loss function that leads to MAP estimators
      for continuous (A) and discrete (B) cases. 47

2.5   Example in 1D of two different imaging systems (ξ_1 and ξ_2) t gener-
      ate two different posteriors. Width of those posteriors is indicative
      of how well the system performs. 62

3.1   The summary of laws (models) used for statistical inferences (dist.
      = distribution). The numbers in the light boxes indicate section
      numbers where the laws are derived and described in details. The
      solid black arrows and gray dotted lines indicate the exact and
      approximate derivations, respectively. 68
3.2   Schematic representation of a voxel with six radioactive nuclei
      shown as balls inside the voxel cube 69
3.3   The line represents the time from t_1 to t. The time is divided into
      small intervals dt. 70
3.4   The discrete binomial distribution of the number of detections from
      a voxel containing $r = 6$ radioactive nuclei. Three types of symbols
      correspond to different detector sensitivities equal to $\alpha = 0.1$, $\alpha =$
      0.5, and $\alpha = 0.9$. Data is represented by symbols (21 in total) and
      lines are added for improved clarity. 74
3.5   Schematic drawing of a single voxel with index $i = 1$ and multiple
      detector elements (four) indexed by $k = 1, \ldots, 4$. 75
3.6   Schematic drawing of a multiple voxel image ($I = 16$) indexed by
      i and multiple detector elements ($K = 4$) indexed by k 77
3.7   Simple tomographic system (STS). The g_1, g_2, and g_3 indicate the
      three detector elements. The voxels are indicated by c_1, c_2, and c_3,
      which correspond to the numbers of events that occurred in voxels
      and were detected. Non-zero values of $\mathbf{y}$ are also specified. 80
3.8   Schematics of the simple tomographic system (STS) as used in Ex-
      ample 3.4. The g_1, g_2, and g_3 indicate the three detector elements.
      The voxels 1, 2, and 3 emitted 5, 1, and 1 detected counts ($\mathbf{c}$) 84
3.9   Comparison of binomial and Poisson distributions for ^{201}Tl (upper)
      and ^{82}Rb (lower). 91
3.10 Schematic drawing of a single voxel (shaded) with index $i = 1$ and
      multiple detector elements (four) indexed by $k = 1, \ldots, 4$. The
      number of decays that were not detected and the number of nuclei
      that did not decay comprise category 0 of multinomial distribution. 92

4.1   500 samples drawn from the bivariate normal distribution with
      $\boldsymbol{\mu} = [5, 5]$ and Σ specified by values $1, -0.5, -0.5$, and 2 are shown
      in the center. Histograms are shown with bin width of 0.5 centered
      at values $0.5i$ where $i \in \mathbb{Z}$ 105
4.2   Total error from MC experiment indicated by bars. The line cor-
      responds to a normal distribution (the marginal of the bivariate
      normal (Equation (4.16)) used in this example) 106

4.3 Monte Carlo estimates of the density values indicated by the points. The error bars correspond to the estimate of MC error using Equation (4.18). The line corresponds to a normal distribution (the marginal of the bivariate normal (Equation (4.16)) used in this example).. 106

4.4 Monte Carlo sampling from interval $[A, B]$ with sampling density $p(f)$. The chance that a MC sample falls within the Δ_k is equal to $p_k = \int_{\Delta_k} p(f)df$ as illustrated... 108

4.5 Generation of samples from $p(g) = 6/(\pi g)^2$. First, one sample x from $\mathcal{U}(0, \pi^2/6)$ is obtained. If the value of $x \leq 1$ $g = 1$ then sample is 1. Otherwise, if $1 < x \leq 1 + \frac{1}{4}$ sample is 2. Otherwise if $1 + \frac{1}{4} < x \leq 1 + \frac{1}{4} + \frac{1}{9}$ sample is 3, etc. 111

4.6 Estimates of values of $\hat{p}(f)$ obtained at 0.1 intervals (connected by line) using Equations (4.39) and (4.40) and $R = 10000$. Both estimates perfectly overlap. 112

4.7 Four diagrams of possible selection schemes for creation of the Markov chain from Example 4.4. 119

4.8 Acceptance frequencies for Markov moves defined in Figure 4.7 (upper/left diagram). The naïve frequencies are defined in Example 4.4. The Metropolis–Hastings frequencies are derived in Section 4.4.4. The "reverse" indicates the reverse moves to ones indicated on x–axis.. 121

4.9 Burn-in times for different starting points 1, 2, 3, or 4 in the phase space... 121

4.10 The value of the first coordinate of the vector $\mathbf{x}$ for the first 1000 steps of Markov chain sampling of multivariate normal distribution described in Example 4.6.. 123

5.1 Atomic notation. 131
5.2 Decay scheme diagram for ^{15}O.. 133
5.3 Decay scheme diagram for ^{133}Xe which is a β^- emitter. There are several prompt gammas (γ_1 through γ_6) that are produced. 134
5.4 Decay scheme diagram for ^{18}F. 136
5.5 Diagram of Compton scattering. Incident photon with energy E_0 is scattered on an outer shell electron. 137
5.6 Diagram of the photoelectric interaction. The energy of the incident photon is transferred to one of the electrons. The electron is ejected from the atom with the kinetic energy equal to the difference between E_0 and the binding energy of the electron. 139
5.7 Photon attenuation. 140
5.8 (A) A point source of photons is located at r_1, r_2 and r_3 distance from three small perfect detectors. (B) The point source is placed in the center of attenuating sphere with radius d. 141

5.9   The CZT is placed in an electric field. Incident photons create electric charge (electron-hole pairs) in the CZT material................. 143

5.10 Scintillation radiation detector.. 144

5.11 Diagram showing basic principles of the PMT. The gray line illustrates the stream of electrons. 145

5.12 (A) Basic operation of the Geiger-mode photodiode. (a) The initial state. Incident photon rapidly raises the current (b) which triggers drop in bias due to use of quenching resistor and the photodiode resets to the initial state. (B) Schematic representation of the array of photodiodes with quenching resistors (microcell). 146

5.13 Voxel representation of imaged volumes... 149

5.14 Event and count timeline. 151

5.15 The voxelized imaging volume (A). An example of ROR (darkened voxels) (B) that corresponds to line. (C) and (D) show RORs that correspond to a plane and a volume... 154

5.16 Gamma camera. 155

5.17 Gamma camera with a collimator and the object (shaded oval). Only gammas numbered 5, 6, and 7 produce a count on the detector.156

5.18 Four most common types of collimator used in nuclear imaging. There are parallel-hole (A), pin-hole (B), converging (C), and diverging (D) collimators. Dotted line shows the path of a detected photon. 157

5.19 Data acquisition in SPECT using two-head scanner. 158

5.20 Schematics of β^+ nuclear decay with indirect production of 511 keV photons.. 160

5.21 Block detector used commonly in PET cameras. 160

5.22 PET camera built from 24 PET block detectors. 161

5.23 Schematics of various detections in PET. (A) true coincidence. (B) Single annihilation photon detection (a single). (C) Scatter coincidence detection (s = scatter location). (D) Random coincidence detection. 163

5.24 Detection of a direct (true) coincidence by two detector blocks (a and b).. 164

5.25 PET and TOF-PET count localization. 165

5.26 In Compton camera, gamma ray emitted in the object is Compton scattered on detector 1 (scatterer) and absorbed in detector 2 (absorber).. 167

5.27 Construction of the ROR for Compton camera................................. 168

5.28 Representation of the voxel or ROI. The FDG tracer is assumed in either of three states: (1) blood plasma, (2) extra-vascular space, or (3) phosphorylated state. 170

5.29 Three-compartment model. Compartment one corresponds to the tracer in the plasma $(C_P(t))$. 171

5.30 The acquisition time is divided into intervals. Data acquired during each interval is used to reconstruct images—one image per one interval. The ROI or multiple ROIs are defined and concentration values in the ROI are determined for each interval (time frame). The thick lines correspond to the average values of continuous function shown as a dashed line. 172

5.31 Conceptual drawing of soft-gamma detector using the Compton camera. Drawing made based on Tajima et al. [102] 177

6.1 Illustration of disjoint subsets $\mathbf{Y_c}$. In this example the set $\mathbf{Y}$ is divided into nine disjoint subsets $\mathbf{Y_{c_1}}, \ldots, \mathbf{Y_{c_9}}$ that provide the *exact cover* of the set $\mathbf{Y}$. 181

6.2 Representation of macroscopic and microscopic states for 3-voxel system. 184

6.3 (A) The Markov chain built from $\mathbf{y}$ vectors. (B) and (C) Markov moves for 3-voxel 6-event model. 186

6.4 Two subsequent states in the OE algorithm Markov chain if the "move" is successful. Squares represent voxels. 189

6.5 Schematics of the simple tomographic system (STS). 194

6.6 Marginalized distributions of $p(c_1|\mathbf{G} = \mathbf{g})$ (left), $p(c_2|\mathbf{G} = \mathbf{g})$ (middle), and $p(c_3|\mathbf{G} = \mathbf{g})$ (right). 196

6.7 Marginalized distributions of $p(d_1|\mathbf{G} = \mathbf{g})$ (left), $p(d_2|\mathbf{G} = \mathbf{g})$ (middle), and $p(d_3|\mathbf{G} = \mathbf{g})$ (right). 196

6.8 Marginalized distributions of $p(d_1, d_2|\mathbf{G} = \mathbf{g})$ (left), $p(d_1, d_3|\mathbf{G} = \mathbf{g})$ (middle), and $p(d_2, d_3|\mathbf{G} = \mathbf{g})$ (right). Clearly, correlations between d_1 and d_2 (more oval shape of the distribution) are much stronger than correlations of quantity d_3 with either d_1 or d_2. 197

6.9 Gamma priors of $\mathbf{d}$ used in the experiment. On the left are gamma priors used for pixel values d_1 and d_2, and on the right the gamma priors used for values of pixel d_3. The 'hi' indicates high confidence in the prior (albeit incorrect) and 'lo' the low confidence in the prior. 199

6.10 Marginalized posterior distributions of $p(d_1|\mathbf{G} = \mathbf{g})$ (left), $p(d_2|\mathbf{G} = \mathbf{g})$ (middle), and $p(d_3|\mathbf{G} = \mathbf{g})$ (right) for gamma priors. Posteriors marked with 'hi' and 'lo' indicate that the posterior was computed from initial high-confidence and low-confidence priors. The simulated true values of $\mathbf{d}$ were equal to 2000, 2000, 10000 for voxels 1, 2, 3, respectively. 200

6.11 Marginalized posterior distributions of $p(d_1|\mathbf{G} = \mathbf{g})$ (left), $p(d_2|\mathbf{G} = \mathbf{g})$ (middle), and $p(d_3|\mathbf{G} = \mathbf{g})$ (right) for Jeffreys prior. Different posteriors correspond to different system matrices (different imagining systems) defined by Equation (6.48) 202

B.1 The set W' is a subset of the set W. Alternatively we can say that W contains W'. 214

List of Tables

1.1 The model: values of conditional probability of OQ g if UQ has value f 16

1.2 Value of the joint probability f and g 23

1.3 Value of the pre-posterior of f conditioned on g where g is assumed to be observed in the experiment (dice rolls) 23

1.4 Definition of the model $p(g|f, \tilde{f})$, the prior $p(f, \tilde{f})$, the pre-posterior $p(f, \tilde{f}|g)$, and the pre-posterior with marginalized NUQ 28

2.1 Values of the pre-posterior of θ conditioned on g where g is assumed to be observed in the experiment (dice rolls) 43

2.2 Values of the posterior expected loss for the decision δ corresponding to pre-posterior in Table 2.1 and the quadratic loss...................... 44

2.3 MMSE estimators with the posterior standard errors......................... 45

2.4 Comparison of the lowest posterior expected loss, MMSE, and MAP estimators 48

2.5 Values of posterior expected losses for composite hypotheses 52

2.6 Generalized Bayesian likelihoods and likelihood ratios 56

2.7 Bayes risk under quadratic loss function computed using Equations (2.32) and (2.33).. 59

2.8 Bayes risk for loaded dice and MSE computed using Equation (2.33) 61

3.1 Probability distribution $p(\mathbf{y}|\mathbf{c})$ calculated using Equation (3.24) for values of $\hat{\alpha}$ and $\mathbf{c}$ specified in Example 3.4 for the STS................. 85

5.1 Properties of scintillator and semiconductor materials used in nuclear imaging 144

5.2 Properties of various photomultipliers [90] 147

5.3 Relationships between activity and radiodensity for common nuclides used in nuclear imaging 150

5.4 Examples of tracers used in nuclear imaging...................................... 175

5.5 Isotopes observed in novae (most apparent in the sky)....................... 176

6.1 Results of the image reconstruction for STS 198

6.2 Posterior $p(H_1|\mathbf{G} = \mathbf{g})$ for various x^a 204

About the Series

The *Series in Medical Physics and Biomedical Engineering* describes the applications of physical sciences, engineering, and mathematics in medicine and clinical research.

The series seeks (but is not restricted to) publications in the following topics:

- Artificial organs
- Assistive technology
- Bioinformatics
- Bioinstrumentation
- Biomaterials
- Biomechanics
- Biomedical engineering
- Clinical engineering
- Imaging
- Implants
- Medical computing and mathematics

- Medical/surgical devices
- Patient monitoring
- Physiological measurement
- Prosthetics
- Radiation protection, health physics, and dosimetry
- Regulatory issues
- Rehabilitation engineering
- Sports medicine
- Systems physiology
- Telemedicine
- Tissue engineering
- Treatment

The *Series in Medical Physics and Biomedical Engineering* is an international series that meets the need for up-to-date texts in this rapidly developing field. Books in the series range in level from introductory graduate textbooks and practical handbooks to more advanced expositions of current research.

The *Series in Medical Physics and Biomedical Engineering* is the official book series of the International Organization for Medical Physics.

The International Organization for Medical Physics

The International Organization for Medical Physics (IOMP) represents over 18,000 medical physicists worldwide and has a membership of 80 national and 6 regional organizations, together with a number of corporate members. Individual medical physicists of all national member organisations are also automatically members.

The mission of IOMP is to advance medical physics practice worldwide by disseminating scientific and technical information, fostering the educational and professional development of medical physics and promoting the highest quality medical physics services for patients.

A World Congress on Medical Physics and Biomedical Engineering is held every three years in cooperation with International Federation for Medical and Biological Engineering (IFMBE) and International Union for Physics and Engineering Sciences in Medicine (IUPESM). A regionally based international conference, the International Congress of Medical Physics (ICMP) is held between world congresses. IOMP also sponsors international conferences, workshops and courses.

The IOMP has several programmes to assist medical physicists in developing countries. The joint IOMP Library Programme supports 75 active libraries in 43 developing countries, and the Used Equipment Programme coordinates equipment donations. The Travel Assistance Programme provides a limited number of grants to enable physicists to attend the world congresses.

IOMP co-sponsors the *Journal of Applied Clinical Medical Physics*. The IOMP publishes, twice a year, an electronic bulletin, *Medical Physics World*. IOMP also publishes e-Zine, an electronic news letter about six times a year. IOMP has an agreement with Taylor & Francis for the publication of the *Medical Physics and Biomedical Engineering* series of textbooks. IOMP members receive a discount.

IOMP collaborates with international organizations, such as the World Health Organisations (WHO), the International Atomic Energy Agency (IAEA) and other international professional bodies such as the International Radiation Protection Association (IRPA) and the International Commission on Radiological Protection (ICRP), to promote the development of medical physics and the safe use of radiation and medical devices.

Guidance on education, training and professional development of medical physicists is issued by IOMP, which is collaborating with other professional organizations in development of a professional certification system for medical physicists that can be implemented on a global basis.

The IOMP website (www.iomp.org) contains information on all the activities of the IOMP, policy statements 1 and 2 and the 'IOMP: Review and Way Forward' which outlines all the activities of IOMP and plans for the future.

Preface

My main reason for writing this book is to introduce Bayesian approaches to analysis of randomness and uncertainty in nuclear imaging. Bayesian methods used in imaging research and presented in many imaging texts do not reflect the Bayesian spirit of obtaining inferences about uncertain phenomena, which is quite unfortunate. Most users who utilize statistical tools in their research consider the probability as the frequency of occurrence of some random phenomena. The word "random" indicates that if an identical experiment is performed repetitively the results may be different and unpredictable. We all have gone through examples of frequentist[1] coin flipping and die rolling way too many times. However, the interpretation of probability as a frequency makes practical applications of drawing inferences quite limited. Anyone who uses computing to analyze experiments and to model uncertainty will certainly encounter serious limitations and complexity of the frequentist techniques. At this point, many of us reach toward Bayesian methods because the Bayesian techniques are overwhelmingly simple to implement to a wide variety of scientific problems.

However, there is a caveat. When we adopt Bayesian methods we often still carry a heavy baggage of misconception of the probability being a frequency. This misapprehension leads to misuse, confusion, and certainly to misinterpretation of results of our analyses. I've heard often conjectures stating that "Bayesian results are biased." For a Bayesian, this statement sounds like "The weather today is yellow"—nonsense! In an effort to explain many of the misconceptions, I introduce Bayesian statistics and present the Bayesian view on uncertainty in the first two chapters of the book. These two chapters are probably the most important because in order to fully embrace Bayesian methods, one needs to purge his or her mind of the concept of probability being a frequency.

Another objective of this book is to introduce Bayesian computational techniques in nuclear imaging. I introduce the statistical model in Chapter 3. Many will find the model unorthodox, as I deviate from the routinely used Poisson statistics in order to derive computationally efficient numerical implementation of Bayesian methods. The Poisson distribution of nuclear imaging hinges on the assumption that the generative process of events is described by an independent event rate. Such a rate is a mathematical construct that helps simplify the description of the decay laws. The gamma radiation, however, is the result of the decay of radionuclei, and I build the statistical model from the ground up based on this simple assumption. Then, I demonstrate how the

---

[1] A frequentist is someone who interprets probability as a frequency of occurrence.

derived decay laws can be approximated by Poisson independent rates and Poisson distribution. Although the goal of Chapter 3 is to provide a statistical basis for algorithms developed in Chapter 6, the theory may also be useful for a better understanding of the counting statistics in general.

With the advent of readily available and inexpensive computing, Monte Carlo methods based on Markov chains became a workhorse for Bayesian computing. They can be used to approximate multi-dimensional distributions (e.g., posteriors) that are otherwise very difficult to characterize. I introduce these techniques in Chapter 4. A short introduction to nuclear imaging and several concepts used in analysis of nuclear imaging data are provided in Chapter 5. The final chapter, Chapter 6, provides derivations of Markov chain algorithms applicable to analysis of nuclear data. It contains demonstrations of calculations of estimators, intervals, Bayes factors, Bayes risk, etc. These examples of Bayesian analysis are provided with the hope that they will inspire readers to use presented methods in problems they face in their work. A sample C++ code that was used in Chapter 6 is provided in Appendix F.

Who should read this book? The book is addressed to a wide spectrum of practitioners of nuclear imaging. This includes seasoned scientists who have not been exposed to Bayesian paradigm as well to students who want to learn Bayesian statistics. I believe that many may benefit from reading Chapters 1 and 2 in understanding of the Bayesian methods in general. My description of the counting statistics in Chapter 3 will also benefit practitioners of nuclear data analysis because they provide complete in-depth derivation of the statistical model of nuclear decay and photon counting. The chapter dedicated to Monte Carlo methods (Chapter 4) and the introductory chapter dedicated to nuclear imaging (Chapter 5) are intended for readers who are not accustomed with basic ideas of Monte Carlo methods and nuclear imaging. These chapters can be skipped by someone familiar with these topics. Chapter 6 is intended for readers looking for alternative methods to nuclear data analysis. The chapter is short and intended to be an inspiration for investigators to discover new ideas and methods of advanced Bayesian data processing in nuclear imaging. My hope is that this chapter will promote new ideas and support development of the field of nuclear imaging data analysis in the future.

I had the privilege and was fortunate to work with many great imaging scientists and I am very grateful to my mentors, collaborators, postdocs, and students who, in one way or another, contributed to this book. I want to acknowledge Anna Celler and Grant Gullberg for early discussions that eventually led to many concepts discussed here. I received many useful and thoughtful comments about this manuscript from Anna Celler, Mellisa Haskell, Joaquín López Herraiz, Sylwia Legowik, Peter Malave, Stephen Moore, and Hamid Sabet, and I would like to thank and acknowledge them for their input and help.

Arkadiusz Sitek

About the Author

Arkadiusz Sitek, Ph.D., is an associate physicist at Massachusetts General Hospital in Boston and an assistant professor at Harvard Medical School. He received his doctorate degree from the University of British Columbia, Canada, and since 2001 has worked as a nuclear imaging scientist in the Lawrence Berkeley National Laboratory, Beth Israel Medical Center, and Brigham and Women's Hospital before joining Massachusetts General Hospital in 2012. He has authored more than 100 scientific journal and proceedings papers, book chapters, and patents, and served as a principal investigator on nuclear imaging research projects. Dr. Sitek is a practitioner of the Bayesian school of thought and a member of the International Society for Bayesian Analysis.

1 Basic statistical concepts

1.1 INTRODUCTION

This chapter and the next chapter are essential for understanding the content of this book. By design, the chapters present statistics from a quite different perspective as usually the statistics is introduced and taught. The theory of statistics is presented from the pure Bayesian perspective where we attempt to make sure that concepts of the classical statistics are not mixed in the exposition of the theory. In our experience, the Bayesian statistics is frequently introduced in image and signal analysis texts as an extension of the classical treatment of probability. The classical treatment of probability is based on the interpretation of probability as the frequency of occurring of some phenomena based on repeated identical trials. The classical approach is often referred to as the *frequentist* statistics. From the Bayesian point of view, the probability describes the strength of beliefs in some propositions. One of the most frequently used terms, the probability distribution, in frequentist statistics means the "histogram" of outcomes of the infinite number of repetitions of some experiment. In Bayesian statistics, the probability distribution quantifies beliefs or in other words measure of uncertainty. Unfortunately, these two concepts of probability, Bayesian and frequentist, are not compatible and cannot be used together in a logically coherent way. What creates confusion is that both approaches are described mathematically by the probability calculus and because of that they can be intermingled and used together which, to us at least, is incomprehensible.

In this book we decided not to introduce classical concepts at all. To help the reader who is accustomed to thinking about the probability as a frequency, we intentionally do not use the term *random variable*. This is because the random variable is strongly associated with the concept of frequency. To avoid any unwanted associations, the term random variable is replaced in this book by the term *quantity*. The classical term *parameter* is not used in this book either. In the classical statistics, parameters describe unknown values and inference about those parameters is obtained in classical statistical procedures. Instead of the term "parameter" the term *quantity* is used as well. Both the "random variable" and the "parameter" are put on the same conceptual level and are referred to as *quantities*. Finally, in the classical statistics the term *data* is used to describe the outcome of experiments. Based on the data, inferences about parameters are made in frequentist statistics. In the Bayesian view utilized here, the term data is another *quantity* which is conceptually the same as quantities corresponding to random variables or quantities corresponding to parameters. For this quantity we relax our naming rule and use interchangeably the data and the quantity to describe outcomes of the experiments.

It may appear that such convention creates confusion because there is a single term "quantity" to describe so many phenomena. There is more to gain than lose as we believe that this naming convention helps considerably with understanding of Bayesian concepts. In order to help differentiating different quantities, we will use adjectives *observable* and *unobservable* added to the term quantity that identify which quantities are revealed in the experiment (correspond to "data" in classical treatment) and which are never revealed (correspond to parameters in classical statistics).

1.2   BEFORE- AND AFTER-THE-EXPERIMENT CONCEPTS

In this chapter, a specific view on processes that involve uncertainty will be considered. The author hopes that the approach will allow to smoothly introduce concepts that are frequently poorly explained or misunderstood. The content of this book is concerned about knowledge of *quantities* that can, or cannot, be observed directly in an experiment. Such quantities will be referred to as observable and unobservable quantities, respectively. Interchangeably, we will refer to knowledge about quantities as beliefs. We will also use uncertainty about the quantity which is the opposite term to knowledge. For *unobservable quantities* (UQs) the true value of the quantity is unknown (uncertain). For example, suppose we are interested in a true weight of some object. This quantity cannot be observed (determined) directly and the true weight is unknown. By unobservable directly we mean that there is no experiment that can reveal the true value of that quantity. The *observable quantities* (OQs) will be those where the true values are revealed by the experiment. For example when weighing an object the reading from the scale is an observable quantity. Obviously, the true weight of the object (unobservable quantity) and the reading from the scale (observable quantity) are two different quantities and are not necessarily equal.

> **Important:** Here an important distinction has to be made. The weight of the object and the result of the measurement are two different quantities. The weight is uncertain before and after the experiment; however, the measurement is uncertain before the experiment (we do not know what the reading on the scale will be), but it is known exactly after the experiment. Therefore the quantity which is the measurement is revealed and known exactly. The true weight remains uncertain.

The quantities that we will be interested in are going to be referred to in this book as the *quantities of interest* (QoIs) which include UQs and OQs. Sometimes quantities that are known will be required to fully describe a problem at hand (when considering the radioactive decay such quantities can be the half-life or decay constant for given radiotracer). These quantities will be referred to as *known quantities* (KQs). The values of all QoIs constitute the objective truth that will be referred to as the "state of nature" (SoN). Obviously the KQs also describe the SoN but since they are known at all stages of the experimentation they are not considered as a part of QoIs. We require that the SoN is defined by at least one QoI. We assume that knowledge of the true SoN implies the knowledge of all true values of QoIs that define it and vice versa. All true values of QoIs, observable and unobservable, define the SoN.

Although some QoIs are not observable, we will be able to make a guess about the true value based on our general knowledge and experience and maybe some experiments that shed some light on the true values of the QoIs that were done in the past. There are two extremes in the amount of information that we can have about a QoI. A perfect knowledge is when we know the true value of the quantity and the least knowledge is when we have no indication which of the possible values of the quantity is the true value.

One way to think about asking how accurate is the information regarding some QoI is to think about a range of possible true values of this quantity. If the number of such values is small, we say that our information is more precise, or better, than information in the case where the number of possible values is larger. In the extreme, for a single possible value, the knowledge is "perfect" and no uncertainty is involved. The knowledge is perfect from the definition for all QoIs that are observable after the experiment performed. If all QoIs are OQs, after the experiment all values are certain, the SoN defined by those quantities is therefore known and statistical description is not necessary.

The goal of any experiment is to improve knowledge about QoIs and the SoN defined by those QoIs. For OQs this improvement is obvious as the true values of those QoIs are simply revealed and the knowledge about them becomes perfect (we know the true values of the QoIs) once the experiment is performed. Sometimes we will refer to those true values of OQ as *experimental data*, *data*, or *observations*. We often will say that the OQ is revealed or realized in the experiment as opposed to hidden, uncertain, or unknown. Based on observable QoIs that are revealed, some additional information about unobservable QoIs will be obtained. This process will be referred to as the *statistical inference*. We deliberately do not use the term *random variable* to describe the QoI, because the word "random" is misleading and makes it difficult to understand the line of reasoning employed in this work. The quantities we refer to as UQ and OQ are deterministic and constant and using the term "random" when referring to them would be confusing. Another deviation from the other authors is that the term *parameter* typically used in the literature to describe some unknown property of the state of nature is not used. The closest correspondence to the classical term "parameter" used in this book is the UQ. We do however place OQs and UQs on the same conceptual level and consider them as quantities that define the SoN.

We consider two stages at which the information about the SoN is evaluated: before-experiment (BE) and after-experiment (AE). When considering the SoN after the experiment (AE), the uncertainty about SoN is described only by the UQs. In the AE stage, the OQs are no longer uncertain and are known; therefore, no uncertainty about them can contribute to uncertainty about the SoN. Just to be sure that there is no misunderstanding, the OQ is the measurement (e.g., reading on the scale) and not the value of the quantity that it attempts to estimate. If the goal of the investigation is to obtain insights about the SoN, it is obvious to consider only the AE stage. However, BE state is also interesting when we do not want to tie conclusions about the SoN to actual observations of OQs, but rather consider all possible observations that can occur. This can be important when our task is to optimize imaging systems in which case we need to consider all possible values of OQs that can be obtained with that imaging system.

Let's consider the following example of the before-experiment (BE) and after-experiment (AE) concepts and a single OQ.

Example 1.1: Single-die roll (1)

The experiment is defined as a simple roll of a six-sided die. The result of the roll is the observable QoI. The SoN is defined by the number obtained in the roll. In the BE state, there are six possible true values of the QoI. The experiment is performed and the number six is obtained. Therefore, the AE state (after the roll) OQs are revealed (realized) so the true value of the QoI (six) is known. In the AE state, there is no uncertainty for this example as the SoN is defined by a single QoI that is known in the AE state.

The same concepts can be illustrated using a more sophisticated example in which the SoN is defined by two QoIs in which one is observable and one is unobservable:

Example 1.2: Radioactive sample (1)

Let's consider a sample of some radioactive material that is expected to create f radioactive decays (total activity) per second. The value of f is unknown and considered as the *unobservable quantity* since it cannot be observed directly. Therefore, per our definition, the value f is a UQ. The experiment is performed that involves observation of g decays from the radioactive sample during some period of time using a detector that registers photons emitted from the sample. The *sensitivity* of the detection is assumed known. The sensitivity is the deterministic constant (KQ) indicating the average of the ratio of the number of detected photons to the number of emitted photons. For simplicity we assume that we have a perfect efficiency and therefore 100% of emissions are registered by the detector. The number of detected counts is a OQ. The SoN (defined by f and g) is uncertain BE and AE; however, it seems that AE we have more information about the SoN as one of the QoIs that defines the SoN is known. Looking slightly ahead, the main idea of statistical inference is that the observation of g counts registered by the detector not only reveals QoI g but also improves the knowledge about the activity f (the UQ); therefore, observations not only reduce the number of unknown QoIs but can also improve the knowledge about the UQs.

The concept of the before and after the experiment conditions introduced in this section is illustrated in Figure 1.1. Before experiment (left column of Figure 1.1) the possible true values of the QoIs f and g are from 0 to ∞. After the experiment is performed (right column of Figure 1.1) the OQ is observed and at this stage it is known and equal to g. The UQ f is still unknown (the true value is uncertain) and for the example presented here the initial range of possible true values remains unchanged (0 to ∞).

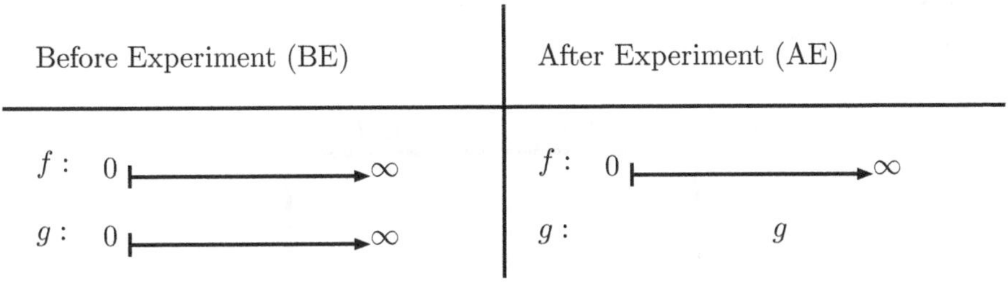

Before Experiment (BE)	After Experiment (AE)
$f:$ 0 ——————→∞	$f:$ 0 ——————→∞
$g:$ 0 ——————→∞	$g:$ g

FIGURE 1.1 Before-experiment and after-experiment concepts.

To summarize, an important concept was introduced that will be used throughout this book. We refer to this concept as the BEAE concept where the BEAE stands for the before-experiment-after-experiment concept. The term experiment indicates a process in which the true values of some quantities are observed (referred to as observable quantities or OQs) which before experiment are unknown. At least one quantity of interest (QoI) should not be observed in the experiment, otherwise no uncertainty would be present in AE (see Example 1.1), all values of QoIs would be known and there would be no uncertainty in the description of the SoN as well since QoIs define the SoN.

Unobservable quantities will be studied in AE state. The uncertainty about UQs will be studied in light of the observed values of OQs. Since OQs are revealed in the experiment, the beliefs about their true values BE are irrelevant.

Although the terms before-experiment and after-experiment suggest that a time series is analyzed, it is not how these two terms should be interpreted. Simply, the term before-experiment indicates that the results of the experiments are unknown but the experiment will be performed and they will become known with certainty. Therefore, in BE state the OQs are uncertain, but their true values are assumed constant. This may sound a bit paradoxical because if we think in terms of time, the experiment has not been performed yet. It is therefore easier to associate the BE state with a stage at which the actual experiment is already performed but simply the results are not yet revealed. This view of the BEAE concept removes the logical paradox. The BEAE concept does not address prediction of future experiments, but rather indicates two stages of knowledge about QoIs considered.

Figure 1.2 presents a summary of all quantities introduced in this section. In this book it is assumed that quantities are numbers and in general can be vectors. All three types of quantities define the state of nature of the system that is being investigated. The true value of observable quantities are revealed in the experiment and the true values of KQ are assumed known in all stages of analysis. The KQs are not part of quantities of interest (QoIs) because the full knowledge about them is available in every state of experimentation. The true values of unobservable quantities are uncertain and the main goal of the statistical analysis is to use available information about the true values of other quantities to reduce uncertainly about the UQs.

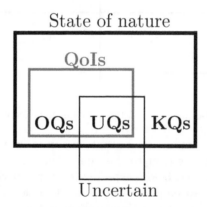

FIGURE 1.2   Observable, unobservable, and known quantities (OQs, UQs, KQs) are identified. UQs are uncertain at the AE stage. Both OQs and UQs are QoIs.

1.3   DEFINITION OF PROBABILITY

At first the probability will be introduced for a single QoI. The term "probability" is used to indicate the strength of our beliefs that some particular value of QoI is true. This definition is very close to the standard use of this term in everyday life. Therefore if we think that some value of QoI is true with a high probability, it indicates a high confidence in this proposition and conversely if the probability that some value of QoI is true is low, our confidence is low.

Suppose we are considering a single QoI in BE state and that the number of possible true values of the QoI is very small. For this case the SoN is defined by a single QoI and therefore probability of the QoI being true is equivalent to the probability of the SoN. The belief that QoI is true is equivalent to the belief that SoN defined by this QoI is true. For example, in a coin toss only two true values are possible: heads and tails. Before the experiment (coin toss) the measure of belief is assigned to each of those values that describes our belief that the result of the toss (true value of QoI) is heads or tails. We will use the term *probability* to refer to those beliefs.

At this point we are ready to define a measure of our beliefs that a value of some QoI is true with more mathematical formality. We assume that for a particular QoI we know all possible true values that the QoI can have. In this book we will use small letters to denote the values of QoIs such as g, f, or y, and corresponding capital letters to denote the set of all possible true values of that QoIs such G, F or Y, respectively. We assume for a moment that the QoIs are scalars and extend the theory to vector quantities in Section 1.8.

As mentioned above, a measure of our beliefs about a given value of a QoI being true is the *probability*. For a given G, any possible QoI $g \in G$ has an assigned measure $p(g)$ (the probability) which is a scalar value from the range 0 to 1. The value of probability 0 about the value of g indicates that the proposition that the g is the true value is impossible and the value of 1 indicates that the true value of QoI is known with certainty and equal to g (this will always be the case for OQs in the AE state).

The following properties of the probability measure are postulated and defined below. Non-trivial extension of the properties of the probability measures to more QoIs will be given in Section 1.4.

1. We postulate that for QoI g where $g \in G$, the probability that the value g is true is described by a number $p(g)$ where $p(g) \geq 0$.
2. We require that the sum over all possible true values of QoI of the probability measure is equal to 1. Therefore $\int_{g \in G} p(g) = 1$.
3. We require that the probability of QoI is either g_1 or g_2 is the sum of probability measures for g_1 and g_2. Mathematically this is denoted by $p(g_1 \cup g_2) = p(g_1) + p(g_2)$.

Example 1.3: Single-die roll (2)

The example of the roll of a die is re-used. Before the experiment (which is the actual roll of a die), assuming the die is fair, we believe that each number that will be revealed in the experiment is equality probable. For this example the true state of nature corresponds to the number that occurs in a die roll. Therefore, denoting by g the QoI indicating the number obtained, the probability a given number is rolled $p(g)$ is $1/6$.

The a priori knowledge or belief is the information about QoI that is available before the experiment is performed. This knowledge is summarized by assigning for each possible value of QoI a measure of belief that this value is the true value. In the example above, the a priori knowledge that a given number would be rolled was $1/6$.

All QoIs can have a prior knowledge assigned to them, but it is meaningful to specify prior knowledge only for QoIs that are unobservable. The reason for this is that a priori knowledge becomes irrelevant for OQs that are revealed in the experiment. Since during the course of planning of the experiment which QoIs will be observed is known, the a priori knowledge for OQ will not be considered.

Note how irrelevant is the fact that we assigned probability $1/6$ to every possible number that can be obtained in a roll once we actually know the number that occurred. Since we know this number in AE condition, all consideration about probability of this number in BE are irrelevant[1].

As introduced in this chapter, two stages of the experiment are identified, BE and AE, and the knowledge (probability, beliefs) about the true value of QoIs may change when considering QoIs in BE and AE states. By convention we will refer to these probabilities as *prior* and *posterior* probabilities at before- and after-experiment stage. We will adhere to this convention throughout this book.

1.3.1   COUNTABLE AND UNCOUNTABLE QUANTITIES

By definition, a *countable QoI* is such that all possible true values of this quantity can be enumerated by integers. Conversely, if the possible true values of a QoI cannot be enumerated by integers, they are labeled as *uncountable QoI*.

To illustrate these countable and uncountable QoIs, consider the two following examples:

Example 1.4: Roll of die and random number generator

The simplest example of a countable QoI is a result of six-sided die roll. In the BE condition the possible values of the QoI are 1, 2, 3, 4, 5, or 6. Therefore the six possible ways can be trivially enumerated by integers 1 through 6 and therefore it is a countable QoI. Another example is an experiment in which a number from range of $[0, 1]$ is selected. For this case in the BE conditions there are an infinite number of values (real number from $[0, 1]$) that cannot be enumerated by integers and therefore the quantity is uncountable. The generation of random number is a common task in computing when using Monte Carlo methods. However, one needs to take into account that real numbers are represented using binary system with limited precision. If double precision is used for

---

[1] In fact the actual prior of OQs can be used to test some assumptions made about the model of the experiment, but this application of the prior of OQ is not discussed in this book. For more on this topic see Berger [8] and Robert [84].

example (64 bits per number) only 2^{64} numbers can be represented. Therefore, when using computers and double precision statistics we actually use countable quantities. For most applications (including medical imaging), this limited precision in representing real number can be ignored, but one needs to be mindful of the limitation of digital machines in representing the uncountable values.

To simplify the notation the symbol $\int_{g \in G}$ is used for both (1) the summation for countable QoI and (2) the integral for uncountable QoI. Which of these two (whether g is countable or uncountable) applies will hopefully be clear from the context. When not obvious it will be specified explicitly if the symbol $\int_{g \in G}$ is a summation or an integral.

We will use the symbol p to indicate the probability or probability density for countable and uncountable QoIs. However, we will use sometimes the term probability for both countable and uncountable QoIs and based on the type of QoI it will be clear probability or probability density is referred to. If the term probability density is used, it will always imply the uncountable QoI.

By $p(g)$ the distribution is indicated, where g can be any of the possible values from the set G. In AE condition some QoIs are observed and at this point their probability distribution is trivial. Only a single value QoI that was observed has a posterior distribution that is non-zero and for all other gs the posterior is zero. The posterior distribution has to obey the normalization condition; therefore, for OQ in AE state, the non-zero posterior for countable and uncountable QoIs is either 1 or the Dirac delta function[2]. Without losing generality, the G will be used to indicate QoIs that are observable (their true value is revealed in the experiment). The following example is used to illustrate the definitions introduced in this section:

Example 1.5: Radioactive sample (2)

Let's revisit the counting experiment (Example 1.2) in which the number of radioactive decays is measured by a perfect sensitivity detector (all radioactive decays are registered). In BE state we have two QoIs f and g where f is a uncountable QoI indicating the amount of radioactivity in the sample. We unrealistically assume that this amount of activity does not change over time and therefore it is assumed constant in time and reflects the average number of emissions per unit time. G is countable QoI and represents the number of decays that will be measured during the experiment. All possible true values of both QoIs are known. The values of f are from a range $[0, \infty]$ and the number of detected radioactive decays g can take integer values $0, 1, \ldots, \infty$. In BE state, we express our beliefs about the true values of the QoIs by the specification of $p(f)$ and $p(g)$ for every possible $f \in F$ and $g \in G$. After the experiment is

---

[2]The Dirac delta function $\delta_d(x)$ is defined such that $\int_{f \in F} \delta(f) = 1$, respectively.

performed and g is measured (observed, realized), the posterior of g is trivial as it is zero for all other than observed number of counts and 1 for the observed number of counts. The prior probability of UQ f, $p(f)$, after the experiment is "updated" to the posterior probability. Interestingly, if we consider another experiment that follows, the posterior from the previous experiment becomes the prior for the new experiment, and is updated again by the data. This type of analysis is called the *sequential analysis* and plays important role in many applications. For more details on sequential analysis refer to Berger [8].

1.4   JOINT AND CONDITIONAL PROBABILITIES

In the preceding sections we considered SoNs that were defined by a single QoI. There, $p(g)$ was the probability that the SoN defined by g was true and similarly $p(f)$ was the probability that SoN defined by f was true. These two different SoNs were considered independently.

Here, we assume that there is only a single SoN defined jointly by f and g. By virtue of this assumption we generalize the probability of such SoN as $p(f, g)$. The comma signifies that we consider a SoN which is defined by particular values f and g. If the SoN is defined by more than one QoI, the probability will be referred to as *joint probability distribution*. For each pair of $\{f, g\}$ that define a possible SoN the probability is assigned. We note the symmetry in the definition. The identical SoN is described by a pair $\{f, g\}$ and by $\{g, f\}$ as there is no significance in the order that we specify the QoIs. This symmetry implies that $p(f, g) = p(g, f)$, so the order of the symbols in notation of joint probabilities is irrelevant.

The axioms that were specified for probabilistic description of SoN described by a single QoI (Section 1.3) apply the same for the SoN described by two (or more as it will be shown in Section 1.8) and therefore:

1. We postulate the probability that the SoN defined by g and f is true is described by a number $p(f, g)$ where $p(f, g) \geq 0$.
2. We require that the sum over the probability of true SoNs (the probability $p(f, g)$) is equal to 1. Therefore $\int_{g \in G} \int_{f \in F} p(f, g) = 1$.
3. We require that the probability of the SoN defined by $\{g_1, f_1\}$ or $\{g_2, f_2\}$ is the sum of probabilities for those two SoNs. Mathematically this is denoted by $p(\{g_1, f_1\} \cup \{g_2, f_2\}) = p(g_1, f_1) + p(g_2, f_2)$.

As defined in the beginning of this section, the true SoN is defined by f and g. If one of those quantities becomes known by obtaining the experimental data g, the uncertainty about the true SoN is manifested only through uncertainty in f. In other words, the probability distribution reflecting our beliefs about the true SoN is the function of only f as the other QoI is known. We indicate this "partial" knowledge of the SoN through the *conditional distribution* $p(f|g)$. We define this distribution in the BE state and therefore only "pretend" that g is known. The conditional distribution can be obtained from the joint distribution simply by extracting values of the joint distribution corresponding to known QoIs and normalizing them by $\int_{f \in F} p(f, g)$. This process is illustrated with Example 1.6.

Example 1.6: Conditional distribution from joint distribution

The concept of the joint probability distribution is illustrated in Figure 1.3. For clarity, we assume that f and g are one-dimensional QoIs and for each pair $\{g, f\}$ the probability is assigned. We first define all possible true values of f and g which is the region $[0, 1]$. An analytical function $p(f, g) = 144(f - 0.5)^2 \times (g - 0.5)^2$ is chosen to represent the joint distribution and plotted in Fig. 1.3(A). We consider a line on 2D plot corresponding to a value $g = 0.3$ (we "simulate" that g is known) and plot values of joint distribution in Figure 1.3(B). Therefore, we "simulate" an experiment in which value 0.3 of QoI g was observed. The analytical form of this distribution is $p(f, g = 0.3) = 144/25(f - 0.5)^2$. Normalization of values of $p(f, g = 0.3)$ by the normalization constant $\int_{f \in [0:1]} p(f, g = 0.3)$ leads to the *conditional probability* which is denoted as $p(f|g = 0.3)$. This notation indicates a conditional probability distribution of QoI f if hypothetically the true value of g is 0.3. The actual shape of the conditional distribution is identical to the joint distribution evaluated at $g = 0.3$ and they differ only by a scaling factor. Although the latter finding is demonstrated on a simple example, it is true in general.

It is easy to demonstrate that all three axioms are obeyed for $p(g|f)$ if they are obeyed for $p(f, g)$. The normalization factor $(\int_{f \in F} p(f, g))$ that was used to obtain the conditional probability we denote as $p(g)$. In fact the normalization factor can be interpreted as a function of g which can be shown to also obey the axioms. The $p(g)$ is the *marginalized* probability distribution obtained from the joint $p(f, g)$ by *marginalization* (integrating out) the other QoI that the joint probability is dependent on (in our example it is f). Sometimes, if the joint distribution has a closed form, the marginalization can be performed analytically (see Examples 1.6 and 1.7). The notation is a little unambiguous as the same symbols were used to describe distribution of g when SoN was described by a single QoI and here where $p(g)$ indicates the marginalized distribution. However, based on the context of whether the SoN is defined by a single QoI or by multiple QoIs would unambiguously indicate the type of distribution that $p(g)$ represents. It follows that if the SoN is defined just by a single QoI the $p(g)$ is the distribution of QoI g in BE condition. If more than one QoIs describe the SoN, the notation $p(g)$ always indicates marginalized distribution.

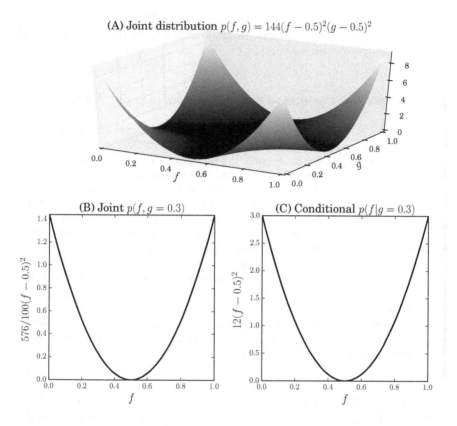

FIGURE 1.3 (A) 2D joint distribution defined for $\{f, g\}$. (B) Values of the joint distribution for QoI $g = 0.3$. (C) Conditional distribution of f conditioned on $g = 0.3$. The shapes of both distributions in (B) and (C) are identical and they differ only by a scaling factor.

Example 1.7: Analytic marginalization

Suppose a joint probability distribution $p(f, g)$ is considered that expresses our beliefs that f and g define the true SoN:

$$p(f, g) = \frac{6}{\pi^2} \frac{1}{g^2} \frac{f^g e^{-f}}{g!} \tag{1.1}$$

The range of possible values of uncountable QoI f and countable QoI g is $F : f \in [0, \infty]$ and $G : g \in [1, 2, ..., \infty]$. We have that:

$$p(g) = \int_{f \in F} p(f, g) = \frac{6}{\pi^2 g^2 g!} \int_0^\infty df\, f^g e^{-f} = \frac{6}{\pi^2 g^2} \tag{1.2}$$

Unfortunately, situations where the analytic marginalization is available is extremely rare in practice and numerical methods are used in order to obtain the numerical approximation of marginal distributions. For example for the case considered here, the marginalization over g yields a sum that cannot be evaluated in a closed-form expression as the sum has no simple mathematical form:

$$p(f) = \int_{g \in G} p(f, g) = \frac{6e^{-f}}{\pi^2} \sum_{g=1}^{\infty} \frac{f^g}{g^2 g!} \tag{1.3}$$

It is left for the reader to check that $p(f, g)$ and $p(g)$ are proper probability distributions and they integrate to 1 over the range of all possible values of f and g.

1.5 STATISTICAL MODEL

There are two motivations that lead to the introduction of conditional distributions. First, once the experiment is performed and OQ are known, there is no point of considering the joint distribution, but rather we "extract" the conditional distribution from the joint, which corresponds to observations, and make inferences based on that. The other reason for conditional distribution is that based on the knowledge of the experiment, we can propose a statistical *model* of the experiment $\mathcal{M}$ by means of conditional distributions.

Before we can define the model, the statistical independence of QoIs needs to be introduced. We define the f *statistically independent* of g when beliefs about f are insensitive to knowledge of true value of g. It follows that statistical independence of f and g implies statistical independence of g and f. Before the experiment, we pretend that g is known and therefore the independence applies to any possible value of f and g.

$$p(f|g) = p(f). \tag{1.4}$$

In other words, our beliefs about f are independent on knowledge of true value of g. The QoIs f and g are *statistically dependent* if $p(f|g) \neq p(f)$. We leave it for the reader to show that if f and g are statistically dependent/independent, then g and f are also statistically dependent/independent. We consider the dependence/independence of f and g in the BE state and this implies that the property is true for any value of QoIs g and f.

We use Example 1.8 to illustrate the concept.

Example 1.8: Conditional distribution from joint distribution – independence

Let's consider an analytical function describing the joint distribution $p(f, g) = 144(f - 0.5)^2 (g - 0.5)^2$. This distribution is in fact an example of a joint distribution of two statistically independent QoIs. This can be demonstrated using simple algebra and by showing that the particular $p(f, g)$ considered in this example implies that Equation (1.4) holds (this is left to the reader). An alternative approach to showing the independence is a graphical demonstration in Figure 1.4 showing that all conditional distributions extracted from the joint distribution are equal.

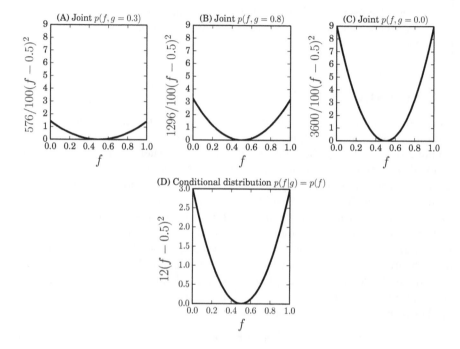

FIGURE 1.4 Three distributions extracted from the joint distribution $p(f, g) = 144(f - 0.5)^2(g - 0.5)^2$ for different values of g equal to (A) 0.3, (B) 0.8, and (C) 0.0. After normalization the same conditional distribution is obtained (D).

So far we used a slightly artificial example of an analytic function representing the joint distribution of two QoIs that are represented by scalar values. To make the concept of statistical dependence/independence more intuitive, a simple experimental model in which two dice are used is considered next. This will also help with the introduction of the concept of the statistical dependence and how it can be used to obtain more information about UQs based on OQs.

Example 1.9: Statistically independent QoIs — dice

Consider rolls of two "fair" dice. The result on only one of the die (say die number two) is revealed to the observer in AE stage. The true values of the numbers obtained in the experiment (the rolls of two dice) are denoted by f and g, respectively, for dice one and two. The possible true values of the QoIs for each roll is 1, 2, 3, 4, 5, or 6. With no other information about the experiment (the model) the result of roll two g (revealed to the observer) does not affect beliefs of the result of the roll one f and vice versa. Therefore, it can be stated that experiment in which the true value g is revealed does change our beliefs about f. This is another way of saying that f and g are statistically independent $(p(f|g) = p(f))$.

It should be quite clear from Example 1.9 that if the QoIs are statistically independent, then in AE the OQ (g) is known and knowledge about UQ (f) is unchanged. In fact, for any statistical problem in which there are some OQs and UQs, the independence would preclude gaining any additional knowledge (reduce uncertainty) about any of the UQs when some statistically independent OQs are observed.

The statistical dependence between observable and unobservable quantities is one of the most important concepts in statistical inference used in this book and it is introduced through the definition of the statistical *model* of the experiment. The model is simply the specification of conditional dependence of OQs and UQs. The model is defined as the conditional distribution where the OQs are conditioned on UQs. Adhering to our convention that an OQ is denoted as g and an UQ denoted as f the model is denoted by $p(g|f)$. If the model is defined, it will imply that f and g are statistically dependent $(p(g|f) \neq p(g))$ otherwise the model would be quite useless for statistical inference (see Example 1.9). The knowledge of the model will be derived from the known physical description of the experiment. When formulating a model, various considerations have to be taken into account. Typically the reality is much more complicated than what can actually be modeled with experiments and the assumed models will simplify the reality in order to be tractable. A trade-off between model complexity and tractability has to be considered in almost every problem.

To better understand the concept of the model, let's consider the extension of the problem with two dice:

Example 1.10: Conditional two-dice roll (1)

Let's consider an experiment in which we have two people. Person A rolls two dice, one after the other. The f describes the result of the first roll and g describes the result of the second roll. Person B ("The observer") observes the result of only the second roll g (since g by convention is used to indicate observable quantity). The value of roll g is the only observable quantity. The result of roll f is unknown to person B. If no other conditions of the experiment are specified (see Example 1.9), the numbers obtained in each of the two rolls are independent, and we will not be able to infer any additional information about f based on g.

However, the experiment is modified such that the person A repeats the second roll g until the number is equal or less than the number obtained in the first roll f and only the result of the last roll is revealed to the observer. By this modification, the statistical dependence is introduced. Based on the description of the experiment, the model of the experiment can be defined. Intuitively, the statistical dependence is obvious since the number on the second roll g will be dependent of the number obtained in the first roll f.

In real systems person A embodies the laws of physics or laws of nature. In this example, the unknown number obtained in the first roll we interpreted as the UQ and the rules governing the process of rolling the second die until the number is equal or less than the number in the first roll are interpreted as the "law of nature."

Just to signal types of problems that will be discussed in this book, the typical question that will be asked is as follows: Having observed the number in the second roll (which is OQ), what can be said about an unobservable quantity of the number obtained in the first roll?

In this book the laws of nature are always described by conditional probabilities $p(g|f)$ (the model) based on the description of the experiment, knowledge of physical principles governing the experimental processes, and logic. For this particular example based on provided description of rules of the experiment, the model $p(g|f)$ is defined in Table 1.1.

TABLE 1.1

The model: values of conditional probability of OQ g if UQ has value f

		\multicolumn{6}{c}{$p(g\|f)$}
	$f \rightarrow$	1	2	3	4	5	6
$g \downarrow$	1	1	1/2	1/3	1/4	1/5	1/6
	2	0	1/2	1/3	1/4	1/5	1/6
	3	0	0	1/3	1/4	1/5	1/6
	4	0	0	0	1/4	1/5	1/6
	5	0	0	0	0	1/5	1/6
	6	0	0	0	0	0	1/6

1.6   LIKELIHOOD

So far several quantities such as joint and conditional distributions were discussed and considered mostly in BE conditions. The model of the experiment summarized by conditional probability $p(g|f)$ where g is the OQ and f is UQ was also introduced. We now move to the AE regime, and in this section we introduce the most important distribution that will be used throughout this book[3], namely, the *likelihood function*. There are two conditions needed to determine the likelihood function: (1) The model has to be known, (2) the data (some OQs used in model definition) need to be observed. Paraphrasing those two conditions, the likelihood function is the "model" $(p(g|f))$ in the AE condition.

The likelihood function (LF) exists only in the AE conditions and exists only when at least one of the QoIs is observed and there is a statistical dependence between the OQs and UQs.

We denote the LF as $p(G = g|f)$ and interchangeably refer to this quantity as the *likelihood function* (LF), the *data likelihood*, or simply as the *likelihood*. The fact that g is observed (AE) is indicated by using the notation $G = g$. The LF for an observed value of g assigns a value of likelihood to every possible $f \in F$ that indicates how likely it is to observe the result $G = g$ if the value of the UQ is f. The LF is therefore a function of the UQ f. The difference between the model $p(g|f)$ and $p(G = g|f)$ is that the first is the distribution defined in the BE and is a function of both g and f, where the latter is a function of f for observed data $G = g$ and defined in AE the condition.

Example 1.11: Conditional two-dice roll (2) — likelihood function

Using the example of the model summarized in Table 1.1 the example of two likelihood functions $p(g = 1|f)$ and $p(g = 5|f)$, for different values of OQ g are shown in Figure 1.5.

The likelihood function is not a probability distribution and therefore we cannot interpret the value of the likelihood function as a measure of probability (belief) of f because in general the likelihood function is not normalized i.e., $\sum_{f \in Y} p(G = g|f)$ is not guaranteed to be 1.

Example 1.12: Conditional two-dice roll (3)

Suppose we now consider the model summarized in Table 1.1 in AE condition

---

[3]The likelihood function is the most important distribution not only here in this book but also in classical statistics [26, 66, 67, 106].

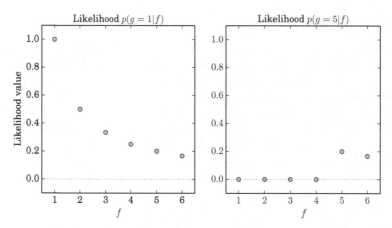

FIGURE 1.5 The likelihood function for the model described in Example 1.10 corresponding to $g = 1$ (left) and $g = 5$ (right). The values in the figure correspond to rows 1 and 5 in Table 1.1.

when $g = 3$ is observed. From the Table 1.1 we extract a row that corresponds to $g = 3$ since these are the only relevant values in AE state. The row has the following values 0, 0, 1/3, 1/4, 1/5, and 1/6. Because these are the values in the AE condition, from the definition they define the likelihood function $p(G = g|f)$. We immediately see that the sum over all possible f does not amount to 1 as the likelihood function is not a probability distribution. Based of the values of the likelihood a likelihood ratio of the form $\Lambda(g; f_1, f_2)$ can be constructed. For example the likelihood ratio $\Lambda(3; 4, 6)/ = 1.5$ indicates that if $g = 3$ is observed it indicates that in the first roll f number 4 is 1.5 times more likely than number 6. We are careful not to use word "more probable" as the likelihood is not a measure of probability. One way to interpret the likelihood is to think about it as a "measure of probability" that comes only from the data. In the next chapter we discuss the likelihood ratio in more details.

We assume that for all problems considered in this book the functional form of the likelihood is known and derived from the model $p(g|f)$. It should be clear that if the model is known then the likelihood is known because from all the distributions $p(g|f)$ defined in the BE state, in the AE state we simply select the distribution corresponding to the observed data g. Interestingly, this function alone can be used to make inferences in classical statistics (e.g., maximum likelihood (ML) methods) but these inferences are quite limited and cannot be easily extended to decision theory. For this reason, the methods that are based solely on likelihood function are not utilized in this book. In general, methods based on data likelihood (e.g., maximum likelihood estimation) do not perform well for systems in which the data contain large amounts of noise and systems with a substantial null space which frequently is the case in nuclear imaging and therefore the ML solution to imaging problems is never used unless some *regularization methods* are employed.

1.7 PRE-POSTERIOR AND POSTERIOR

Throughout this book all decisions about UQ f will be based on *posterior* distributions $p(f|G = g)$. This distribution is defined in AE condition as indicated by the fact that the OQ g is known ($G = g$). Earlier for the case of single QoI, we noted that our general approach to reasoning is that we start with some initial beliefs about a true value of f which we described by the prior distribution $p(f)$. After the experiment is performed and some OQ (g) is observed, the original beliefs are modified to reflect the additional information that is obtained by revealing the true value of the OQ g. Here we describe it mathematically by defining the posterior which is the original beliefs $p(f)$ updated by experiment outcome g. From the definition, the posterior is the probability distribution on f when g is true (observed).

The conditional distribution of UQ conditioned on OQ can also be defined in the BE state. For this case we simply assume that if g is observed it would result in the posterior $p(f|g)$. Because we only assume that g is observed we will refer to the $p(f|g)$ as the *pre-posterior*. There is only one posterior distribution $p(f|G = g)$ but there are many pre-posterior distributions and the number of pre-posteriors is the same as the number of possible true values of g. There are two ways to obtain the posterior distribution which we will discuss below.

1.7.1 REDUCTION OF PRE-POSTERIOR TO POSTERIOR

The most straightforward approach to obtain the posterior is when the pre-posteriors $p(f|g)$ are known. The experiment reveals the true value of g and from the family of pre-posteriors the one is selected that corresponds to g which was observed. The pre-posterior is "reduced" to posterior once the g is observed:

$$p(f|g) \xrightarrow{\;g \text{ is observed}\;} p(f|G = g). \tag{1.5}$$

1.7.2 POSTERIOR THROUGH BAYES THEOREM

In real-world scenarios the direct knowledge of the pre-posterior will not be available and the posterior will be obtained through the Bayes theorem. As explained in Section 1.4 if the joint distribution is known in BE, it is quite straightforward to determine the pre-posterior and once the pre-posterior is known finding posterior in AE state is trivial (see Section 1.7.1). Unfortunately the joint distribution is typically unknown. However, the joint distribution can be approximated using

$$p(f, g) = p(g|f)p(f) \tag{1.6}$$

where on the right-hand side we have a "model" $p(g|f)$ that we assumed is known and the $p(f)$ which is unknown in the large majority of cases. However, since $p(f)$ can be interpreted as the current knowledge about UQ f an educated guess about this distribution based on the current knowledge can be made and joint distribution can be approximated using Equation (1.6). We come back to the problem of the selection of $p(f)$ later in this section. Assuming that the joint probability is known (or approximated) obtaining the pre-posterior (or posterior) directly follows from Equation (1.6) and the fact that if the joint distribution is known, the posterior can be obtained by selecting values of the joint corresponding to the "hypothetically" observed g and then normalizing those values to 1. Using this approach the pre-posterior is readily calculated as follows:

$$p(f|g) = \frac{p(f,g)}{\int_{f' \in F} p(f',g)} = \frac{p(g|f)p(f)}{\int_{f' \in F} p(g|f')p(f')}. \tag{1.7}$$

Often in the literature the normalization term $\int_{f \in F} p(f,g)$ is abbreviated to $p(g)$. The posterior is derived from the pre-posterior replacing a value of g with the actual measurement $G = g$ and the following is obtained that we refer to as the *Bayes Theorem*.

$$p(f|G = g) = \frac{p(G = g|f)p(f)}{\int_{f' \in F} p(G = g|f')p(f')} \tag{1.8}$$

where $p(G = g|f)$ is the likelihood and $p(f)$ is the guess about the value UQ f. The $p(f)$ can also be interpreted as the current state of knowledge about the value of f. It is customary to refer to this distribution as the *prior* to indicate that it represents the knowledge about some QoI prior to the experiment.

The above equation is known as the *Bayes Theorem* named after Thomas Bayes who suggested using the above equation to update beliefs based on data. The theorem gained popularity after it was rediscovered by Pierre-Simon Laplace and published in 1812 [63].

1.7.3 PRIOR SELECTION

One of the most frequent critiques of the use of Bayes Theorem in science is that objectivity of the analysis using posterior requires the accurate knowledge of the prior which is almost never available. It follows that if the guess about $p(f)$ is inaccurate the resulting posterior may be inaccurate as well. We fully agree with these arguments; however, we would like the reader to consider the following three points:

Complete objectivity is an illusion. Drawing statistical inference without the use of Bayes Theorem relies on the model and the resulting likelihood after the experiment is performed. This, however, relies on the assumption that the model is correct. When complex biological processes are involved this will never be the case and some level of approximation when constructing a model will have to be used. Therefore, there is some level of subjectivity used when a model is assumed. Therefore, so-called "objective" analysis frequently lacks objectivity as well because the model is never exact.

Subjectivity is unavoidable in decision making. We should not forget that the endpoint of any analysis in medical imaging or for that matter in medicine is not the posterior or likelihood but rather the decision. Assuming that statistical analysis is performed objectively the decision (disease present or not, optimal course of therapy, etc.) still has to be made and this decision will require a subjective judgment. The definition of prior should be interpreted as an attempt to quantify the use of subjective knowledge, and the use of subjective knowledge is unavoidable.

Bayesian analysis describes knowledge gain. The most importantly, the application of the Bayes Theorem should be interpreted as an update of the current beliefs (reflected by the prior $p(f)$) by the experimental data leading to the posterior $p(f|G = g)$. In this interpretation the correctness of the prior is quite irrelevant with respect to the correctness of the theory and logic of the approach. If the prior is inaccurate, it simply represents inaccurate beliefs that hopefully are improved by the results of the experiment. The correctness of the prior beliefs is not prerequisite for correctness of the analysis. While using wrong assumptions (the prior) may result in wrong conclusion (the posterior), there is no logical inconsistency in this process. The main idea of the analysis is that the beliefs are improved by the data and if the prior is "wrong" the posterior will be "less wrong."

There is a substantial research in the field investigating various approaches to prior selection. In this book we interpret the prior as the current (before experiment) beliefs about the UQs. The prior expresses the level of uncertainty about the true value of the UQ. However, we will also frequently use so-called *uninformative* prior which is attempt to express the belief that we know little about the true values of UQs. For example, one could assign the same prior probability to every possible value of the UQ (*flat prior*) naively believing that this assignment expresses the lack of knowledge. This approach is appealing from the point of view of ease of the implementation, but in fact it is incorrect and does not express our actual beliefs. The flat prior, in fact, implements prior beliefs that all possible true values of UQ have the same probability, which is a quite specific belief about the reality. By doing this we hope that information contained in the likelihood overwhelms incorrect beliefs in the prior. The benefits of the ease of use of flat prior have to be weighted against the possible inaccuracy in the posterior.

A interesting approach to the selection of objective prior (not requiring a subjective judgment) was introduced by Bernardo [10], Ye and Berger [113]. In this approach the selection of the prior is determined by the model in BE stage, and maximizes the expectation of the gain in information (information defined in the Shannon sense [91]) that will be brought by the experiment. Therefore, the prior is used merely as a tool to obtain a posterior, and since the process is deterministic, the statistical analysis that leads to posteriors based on reference priors does not require any subjective judgments.

Another type of priors that we would like to mention is the *entropy priors* championed by Jaynes [47]. The idea is to maximize the entropy of the prior distribution in order to minimize the amount of information that the prior contains (we cover this topic in more detail in Section 6.1.2).

1.7.4 EXAMPLES

To help understand the distributions introduced in previous sections (joint distribution, likelihood, pre-posterior, and posterior), the following two examples briefly introduced in previous sections are used.

Example 1.13: Conditional two–dice roll (4)

The example of the conditional dice roll is used in which person A rolls the first die (f) and then rolls the second die g until number is equal or less than on the first. Only the final number obtained on the second die g is revealed for the observer. Therefore the true values of f and g (the true SoN) are known only to person A and the observer knows only the true value of g in the AE state.

First, we characterize the problem in BE condition. For both QoIs f (the value of first roll) and g (the value of second roll) there are six possible true values of f and g corresponding to the numbers 1 through 6. We assume that the dice are "fair" and therefore assign a priori probabilities for each value equal to $1/6$. This is to ensure that the distribution is normalized to 1. Interestingly, for this case the priors $p(f)$ and $p(g)$ that express the beliefs that the dice are fair are the same as the non-informative flat prior that assigns equal prior probability to every possible true value of the QoI. This is purely coincidental and only infrequently a flat prior will express our exact beliefs.

In order to determine the pre-posterior and posterior for this example, the joint distribution of $p(f, g)$ is determined first. This can be done using Equation (1.6). The statistical model is summarized in Table 1.1. Table 1.2 presents the calculated joint probability. The marginal values of the probability of obtaining g (Table 1.2) calculated as $\int_{f \in F} p(g|f)p(f)$ are presented. The joint distribution summed over all possible SoNs (pairs $\{f, g\}$) is equal to 1 which can be verified in Table 1.2:

Since the joint distribution is specified, the value of the pre-posteriors can be found by normalizing values in each row of Table 1.2 to 1. This is done by dividing values in each row by the corresponding value of $p(g)$ shown in the last column in Table 1.2. The result of this division is shown in Table 1.3.

Having determined the distribution of the pre-posterior, in AE (Table 1.3), the posterior can correspond to any row in the table. The selection of an appropriate row depends on the observed true number obtained in the second roll g. For example, if $g = 2$ is observed, the probability of the value on the first roll f being 2 is three times larger than that of being 6.

TABLE 1.2

Value of the joint probability f and g

				$p(g, f)$			$p(g) = \sum_{f \in F} p(g, f)$
$f \rightarrow$	**1**	**2**	**3**	**4**	**5**	**6**	
$g \downarrow$ **1**	1/6	1/12	1/18	1/24	1/30	1/36	147/360
2	0	1/12	1/18	1/24	1/30	1/36	87/360
3	0	0	1/18	1/24	1/30	1/36	57/360
4	0	0	0	1/24	1/30	1/36	37/360
5	0	0	0	0	1/30	1/36	22/360
6	0	0	0	0	0	1/36	10/360

TABLE 1.3

Value of the pre-posterior of f conditioned on g where g is assumed to be observed in the experiment (dice rolls)

				$p(f|g)$		
$f \rightarrow$	**1**	**2**	**3**	**4**	**5**	**6**
$g \downarrow$ **1**	60/147	30/147	20/147	15/147	12/147	10/147
2	0	30/87	20/87	15/87	12/87	10/87
3	0	0	20/57	15/57	12/57	10/57
4	0	0	0	15/37	12/37	10/37
5	0	0	0	0	12/22	10/22
6	0	0	0	0	0	1

1.7.5   DESIGNS OF EXPERIMENTS

Obtaining statistical inferences about UQ based on experimental data and prior beliefs will be done using the following. The experiments are considered as means of improvement of the information about the UQ. In BE state the knowledge about UQs is described by the prior and in AE state by the posterior. Often, the specification of the prior will be very difficult and a pragmatic approach will be followed compromising between the accuracy in formulation of the prior with the ease of implementation. The goal will always be to obtain the pre-posterior or posterior and based on these distributions make decisions about a problem at hand. The following defines the experimental design that will be followed for every problem discussed in this book:

1. Define BE and AE conditions, identify QoIs that are unobservable (UQ) and QoI that we can measure (obtaining their true value) which are statistically dependent on UQs.
2. For each of QoIs specify the range of possible true values.
3. Based on the description of the experiment, formulate the model of the experiment defined by a conditional distribution.
4. For each of the UQs specify the initial beliefs (prior) in the form of the prior distribution.
5. For UQs determine the pre-posterior distribution based on the model and the prior using Bayes Theorem (Section 1.7). If decisions are based on the actual measurements of OQs determine the posterior.
6. Make a decision based on posterior or pre-posterior depending on a problem at hand.

In this chapter we cover steps 1 through 5, and in the next chapter we will describe approaches to decision making (point 6) based on the posteriors and pre-posteriors.

The steps 1 through 5 define the Bayesian analysis that will be used in all problems that are discussed in this book. The Bayesian analysis of any problem ends with providing the posterior or pre-posterior which contains the complete information about the UQs. This sometimes will not be sufficient in practical applications. For example, we doubt that providing physicians with a million-dimensional posterior distribution would be received enthusiastically. Therefore, the posterior will be summarized in one way or another to provide easily digestible information for the decision maker. For example, the image of the distribution of the radiotracer will be a much more reasonable result provided to physicians than the posterior distribution function. The formation of the image from the posterior is a decision problem, as a decision needs to be made about which possible true values of the UQs best represent the reality. The word "best" used in the previous sentence is undefined at this point and exact definition will be given in the next chapter discussing the decision making.

So far, we have introduced four key distributions pictured in Figure 1.6. Only two of those distributions will be used in this book for decision making, the pre-posterior and the posterior, covered in Chapter 2.

Before Experiment (BE)	After Experiment (AE)
$p(g\|f)$ Model	$p(G = g\|f)$ Likelihood
$p(f\|g)$ Pre-posterior	$p(f\|G = g)$ Posterior

FIGURE 1.6 Distributions introduced in Chapter 1. It is assumed that OQ is g and UQ is f.

1.8   EXTENSION TO MULTI–DIMENSIONS

Up to this point, either a single QoI or a pair of single-dimensional QoIs f and g were considered. In real world applications many more QoIs will be used to characterize a problem. The case of multi-dimensional QoIs is a straightforward extension of the presented theory. To simplify the notation and make it clearer in most cases we will adhere to the convention that all QoIs are divided into two groups: UQs and OQs. We will use vector notation indicating UQs as $\mathbf{f}$ and OQs as $\mathbf{g}$ where I and K indicate the number of elements in those vectors, respectively. Bold non-italic small-letter font will always indicate vectors.

The probability distribution $p(\mathbf{f})$ is defined as the joint probability distribution of components of the vector $\mathbf{f}$:

$$p(\mathbf{f}) = p(f_1, f_2, \dots, f_I) \tag{1.9}$$

Similarly, conditional probabilities $p(\mathbf{f}|\mathbf{g})$ are defined as the joint probability of elements of $\mathbf{f}$ conditioned on the joint probability of elements of vector $\mathbf{g}$ as:

$$p(\mathbf{f}|\mathbf{g}) = p(f_1, f_2, \dots, f_I | g_1, g_2, \dots, g_K) \tag{1.10}$$

Sometimes the notation will be used when more than two symbols will be used to indicate the distributions. For example $p(\mathbf{f}, \mathbf{g}, \mathbf{y})$ where $\mathbf{y}$ is a vector with J QoIs is a joint probability distribution of all elements of vectors $\mathbf{f}$, $\mathbf{g}$, $\mathbf{y}$:

$$p(\mathbf{f}, \mathbf{g}, \mathbf{y}) = p(f_1, f_2, \dots, f_I, g_1, g_2, \dots, g_K, y_1, y_2, \dots, y_J) \tag{1.11}$$

All considerations from previous sections about two scalar QoIs are easily transferable to more than two vector QoIs.

In the following two sections we present rules that will allow transformations of the joint and conditional probabilities of multi-dimensional probability distributions of the QoIs.

1.8.1 CHAIN RULE AND MARGINALIZATION

The chain rule allows expressing the joint probability (e.g., probability distribution of vector QoI) and is the generalization of Equation (1.6),

$$p(\underbrace{f_1}, \underbrace{f_2, f_3, \ldots, f_I}) = p(f_1|f_2, f_3, \ldots, f_I)p(f_2, f_3, \ldots, f_I) \qquad (1.12)$$

where underbraces indicate two probability distributions: probability distribution of f_1 and joint probability distribution $f_2, \ldots, f_I$. Applying the above $I - 1$ additional times the original joint distribution $p(f_1, \ldots, f_I)$ can be expressed as a product:

$$p(f_1, f_2, f_3, \ldots, f_I) = p(f_1|f_2, f_3, \ldots, f_I)p(f_2|f_3, \ldots, f_I) \cdots p(f_{I-1}|f_I)p(f_I) \qquad (1.13)$$

To marginalize multi-dimensional distributions we apply similar rules as in the two one-dimensional distributions a shown in Example 1.13 for two dimensions.

$$p(\mathbf{f}) = \int_{\mathbf{g} \in \mathbf{G}} p(\mathbf{f}, \mathbf{g}) \qquad (1.14)$$

where by $\mathbf{G}$ we indicate all possible values of QoIs $\mathbf{g}$. If we are interested in the marginalized density of a single QoI which is an element of $\mathbf{f}$ the following applies

$$p(f_1) = \int_{\mathbf{g} \in \mathbf{G}} \int_{f_2, \ldots, f_I \in F_2 \ldots F_I} p(\mathbf{f}, \mathbf{g}). \qquad (1.15)$$

> **Example 1.14: Application of chain rule, marginalization, and Bayes Theorem**
>
> Suppose we want to express the conditional probability distribution of a single element of vector $\mathbf{f}$ conditioned on a single element of $\mathbf{g}$ given conditional $p(\mathbf{g}|\mathbf{f})$. The chain rule, marginalization, and Bayes theorem are sufficient for this task:
>
> It is obtained by
>
> $$p(f_1|g_1) = \frac{1}{p(g_1)} \int_{f_2, \ldots, f_I \in F_2, \ldots, F_I} \int_{g_2, \ldots, g_K \in G_2, \ldots, G_K} p(\mathbf{g}|\mathbf{f})p(\mathbf{f}) \qquad (1.16)$$
>
> The distributions $p(\mathbf{f})$ and $p(g_1)$ were required in order to accomplish the task. Distribution $p(g_1)$ can be obtained from $p(\mathbf{g})$ through marginalization.

In order to express the Bayes Theorem and relations between the model and pre-posterior (or likelihood and posterior) using multi-dimensional QoIs, the straightforward generalization of two one-dimensional QoIs is used, and the chain rule described in the previous section,

$$p(\mathbf{f}|\mathbf{g}) = \frac{p(\mathbf{g}|\mathbf{f})p(\mathbf{f})}{p(\mathbf{g})} \qquad (1.17)$$

1.8.2 NUISANCE QUANTITIES

Sometimes when considering many UQs, some of them will not be of direct interest. Therefore their value will influence the posterior and make it a higher dimensional than if the posterior is dependent only on UQs that are of direct interest. Those quantities that are not of direct interest will be referred to as *nuisance QoIs*. Since nuisance QoIs are always UQs they will be denoted as NUQs (N+UQs). In the formulation of the statistical analysis used in this book, nuisance quantities are handled with relative ease. It is done by first determining the pre-posterior or posterior distributions using methodology provided in this chapter and then by marginalizing the NUQs. This can be summarized in the following equations and illustrated by Example 1.15.

Suppose that the vector of NUQs is indicated by $\tilde{\mathbf{f}}$ and the posterior of all unobservable QoIs is indicated by $p(\mathbf{f}, \tilde{\mathbf{f}}|\mathbf{G} = \mathbf{g})$, then the posterior of UQ is obtained by marginalization:

$$p(\mathbf{f}|\mathbf{G} = \mathbf{g}) = \int_{\tilde{\mathbf{f}} \in \tilde{\mathbf{F}}} p(\mathbf{f}, \tilde{\mathbf{f}}|\mathbf{G} = \mathbf{g}) \qquad (1.18)$$

Similar marginalization of NUQs can be used to obtain pre-posterior of the UQ.

Example 1.15: Three scalar QoIs — marginalization of nuisance QoIs

Suppose we consider a simple model of the experiment with three scalar QoI: the OQ g, the UQ f, and the NUQ $\tilde{f}$. Each of the QoIs has only two possible true values of either 0 or 1. The model and the prior are defined in Table 1.4

From Table 1.4 we can formulate the initial beliefs about f and $\tilde{f}$ by marginalizing the prior $p(f, \tilde{f})$ and obtain $p(f = 0) = 0.80$ and $p(f = 1) = 0.20$. Similarly the marginalized prior of NUQ $p(\tilde{f} = 0) = 0.40$ and $p(\tilde{f} = 1) = 0.60$. We also note that the joint prior $p(f, \tilde{f})$ as defined above indicates that f and $\tilde{f}$ are statistically dependent as the $p(f, \tilde{f}) \neq p(f)p(\tilde{f})$ which indicates statistical dependence based on definition in Equation (1.4) because

$$p(f|\tilde{f}) = \frac{p(f, \tilde{f})}{p(\tilde{f})} \neq \frac{p(f)p(\tilde{f})}{p(\tilde{f})} = p(f) \qquad (1.19)$$

and therefore $p(f|\tilde{f}) \neq p(f)$.

TABLE 1.4

Definition of the model $p(g|f, \tilde{f})$, the prior $p(f, \tilde{f})$, the pre-posterior $p(f, \tilde{f}|g)$, and the pre-posterior with marginalized NUQ

$p(g|f, \tilde{f})$ [model]

$\{f, \tilde{f}\} \rightarrow$	0,0	0,1	1,0	1,1
$g \downarrow$ **0**	0.00	0.10	0.30	0.50
1	1.00	0.90	0.70	0.50

$p(f, \tilde{f})$ [prior]

$\{f, \tilde{f}\} \rightarrow$	0,0	0,1	1,0	1,1
	0.30	0.50	0.10	0.10

$p(f, \tilde{f}|g)$ [pre-posterior]

$\{f, \tilde{f}\} \rightarrow$	0,0	0,1	1,0	1,1
$g \downarrow$ **0**	0.00	0.38	0.23	0.38
1	0.34	0.52	0.08	0.06

$p(f|g)^a$ [marginalized pre-posterior]

$f \rightarrow$	0	1
$g \downarrow$ **0**	0.38	0.62
1	0.86	0.14

Note: The values are given with the precision of two decimal places.
[a] The values of $p(f|g)$ are obtained from $p(f, \tilde{f}|g)$ by adding the first two and the last two columns. For this simple example this addition corresponds to marginalization.

Example 1.16: Wire-cube of joint probabilities

In this example the data from the previous Example 1.15 is re-used. The idea is to demonstrate using an illustrative model of wire-cube of joint probabilities (Figure 1.7(A)) that the joint-distribution contains all statistical information about the problem at hand and other distributions can be derived from the joint distribution.

In Figure 1.7, only a few examples are given which demonstrate how to use the wire-cube to obtain other distributions in BE and AE states. However, all distributions can be derived with ease. Although only three dimensions are used and only two possible true values for each QoI, the wire-cube model is correct in any number of dimensions and any number of possible true values for QoIs. Obviously it would be impossible to represent graphically higher dimensional wire-cubes.

In Figure 1.7(B), the wire-cube represents the model (conditional probability) of future observation g conditioned on values of f and $\tilde{f}$. The values of the model are obtained from the joint probability (from the wire-cube Figure 1.7(A)) by selecting the corners in Figure 1.7(A) that corresponds to the same pair of values of $\{f, \tilde{f}\}$ and normalizing them to 1. This is indicated in Figure 1.7(B) by connecting those corners by a thick line. The representation of pre-posterior $p(f, \tilde{f}|g)$ in Figure 1.7(C) is obtained from the joint Figure 1.7(A) by identifying corners that correspond to the same values of the measurement g and normalizing them to 1. Those corners are connected by a thick line shown in Figure 1.7(C). The wire-cube in Figure 1.7(C) represents the pre-posteriors and if the data is measured in AE (either $g = 0$ or $g = 1$) one of the connected thick-line squares corresponding to observed g will become the posterior. Figure 1.7(D) illustrates the posterior $p(f|g)$ which is the result of marginalization of $p(f, \tilde{f}|g)$, which simply adds values of $p(f, \tilde{f}|g)$ in Figure 1.7(C) along $\tilde{f}$ axis. Figuratively speaking, the cube is squeezed and dimensionality of the distribution reduced.

Based on this example as an exercise, the reader may try to obtain various likelihoods or quantities as $p(f)$ etc.

1.9 UNCONDITIONAL AND CONDITIONAL INDEPENDENCE

The last idea introduced in this chapter could be one of the most important concepts for the design of efficient computational algorithms discussed in Chapter 3 and Chapter 6. At the same time, it is one of the hardest to properly understand and gain intuition about. The simple independence of some QoIs f and y was already defined in Section 1.5 and the usual mathematical property indicating this independence is

$$p(f, y) = p(f)p(y) \tag{1.20}$$

with equivalent formulations $p(y|f) = p(y)$ and due to symmetry $p(f|y) = p(f)$. It was explained that independence of y and f indicates that the knowledge of the true value of f does not change uncertainty about y.

If another QoI g is introduced, the joint of three QoIs is defined as $p(f, y, g)$. The joint can be marginalized over f to $p(g, y)$ and after the marginalization the unconditional independence (note that we use the term *unconditional independence* instead of simple independence) is considered again. If the marginalized joint of g and y is independent ($p(g, y) = p(g)p(y)$), then the g and y are *unconditionally independent*. Unconditional independence is defined in situations where at least three QoIs are considered and all QoIs other than g and y are marginalized. Once obtained, the independence for marginalized joint $p(g, y)$ is assessed using standard methods (e.g., Equation (1.20)). The difference between statements that g and y are independent or unconditionally independent is that in the first case the SoN is defined just by two QoIs, whereas unconditional independence is used when SoN is defined by more than two QoIs and other quantities are marginalized.

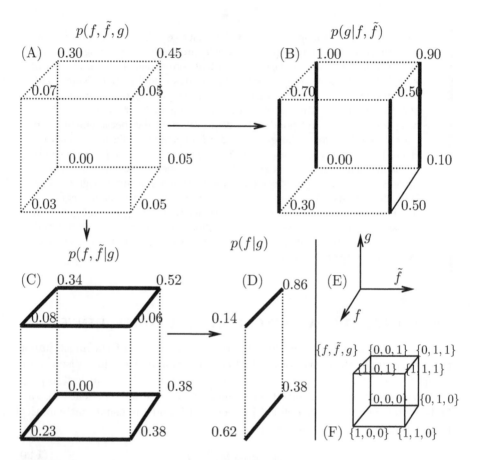

FIGURE 1.7 (A) The wire-cube represents the joint probability of three QoIs f, $\tilde{f}$, and g. There are only two possible true values for each QoI and each corner of the cube correspond to different values of f, $\tilde{f}$, and g, which are shown in (F). The values of each QoI are either 0 or 1. (B), (C), (D) show some conditionals and marginalizations described in detail in Example 1.16.

Consider a *conditional joint* $p(g, y|f)$ which is a joint distribution of g and y conditioned on f which is another way of saying the we consider joint distribution of g and y assuming that the true value of f is known.

Therefore, if the value of f is known we define conditional independence (conditioned on the knowledge of f). Similar to Equation (1.20) the *conditional independence* of g and y given f can be mathematically summarized by

$$p(g, y|f) = p(g|f)p(y|f). \tag{1.21}$$

The above equation has to be true for all $f \in F$ for g and y to be considered conditionally independent given f. Unconditional independence of g and y $(p(g,y) = p(g)p(y))$ does not imply they are also conditionally independent (when the true value f is known) and vice versa.

The definitions are somewhat abstract and therefore we now put these ideas in some more intuitive context.

> **Notation** First let's introduce a symbol $\perp$ which indicates the independence and therefore $g \perp y$ is read as g is independent of y. If g and y are dependent, that will be indicated simply by 'not $g \perp y$'. The conditional independence and dependence of g and y given f are indicated by $g \perp y|f$ and 'not $g \perp y|f$', respectively.

Using these conventions let's consider the following four scenarios:

1. (not $g \perp y$) and (not $g \perp y|f$)

   The QoIs g and y are dependent without any information about f (not $g \perp y$) and with knowledge of true value f (not $g \perp y|f$). This indicates that if either g or y is known, that would change the knowledge about uncertain true value of y or g regardless of whether any knowledge about true value of f is available. However, this does not mean that the change in knowledge will be the same in cases when we have and we do not have information about the true value of f.

2. ($g \perp y$) and (not $g \perp y|f$)

   In this situation without the knowledge of true value of f the g and y are independent and no gain in reducing the uncertainty in g or y can be obtained if the true value of y or g becomes known. Upon knowing the true value of f (more information available) the g and y become dependent. To illustrate this, let's consider the following: Consider two people A and B. The probability of A getting a lung cancer and probability of B getting a lung cancer in the next 10 years without any other information seem to be independent. However, upon obtaining the information that they both smoke, the probabilities are no longer independent, as they both have a higher chance of getting the lung cancer. If the information that is obtained is that person A smokes and person B does not, it makes person A more likely to have a cancer than person B in which case dependence is introduced again, etc.

3. (not $g \perp y$) and ($g \perp y|f$)

   This situation applies to imaging and is discussed in Section 3.3.3. For this case, the unconditional dependence is "removed" if the true value of f is known. Because this case is important for the applications discussed in this book, it is illustrated with Examples 1.17 and 1.18.

4. ($g \perp y$) and ($g \perp y|f$)

   The final case is in a sense the least interesting out of 4 cases listed here. It states that no information can be gained about QoIs g or y

upon knowledge of true value of y or g without or with the knowledge of true value of f.

Example 1.17: Three-dice roll (1) — conditional independence

The conditional example with two-dice roll (Example 1.10) is somewhat modified and three dice are used. Here, a person A rolls die 1 and notes the result f. Then, the same person A rolls die 2 until the number is less than or equal to the number obtained in roll 1 and the result of the last roll becomes g. He repeats the last steps with die 3 obtaining y.

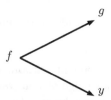

Only the result of g and y are made known to person B who analyzes the problem. Suppose that first we want to establish if there is unconditional independence between g and y. Intuitively, we suspect that there is a dependence because if, for example, $g = 1$ it makes it possible that the unknown f was also 1 in which case y must be 1 as well because of the description of the experiment. It seems that if g is 1 it makes it more probable that $y = 1$ than any other number 2 through 6 compared to a case when no dependence is considered.

In order to verify this let's construct the marginalized distributions $p(g, y)$, $p(g)$, and $p(y)$ and then confirm that $p(g, y) \neq p(g)p(y)$. In doing so we can also confirm our intuition that $p(f = 1|g = 1) > p(f = 1)p(g = 1)$. First, we determine the joint $p(g, y)$ by $p(g, y) = \sum_{f=1}^{6} p(g, y|f)p(f)$. The $p(f)$ is known and equal to $1/6$. The $p(g, y|f)$ can be easily deducted from the description of the experiment and for example for $f = 3$ it is:

	$y = 1$	$y = 2$	$y = 3$	$y = 4$	$y = 5$	$y = 6$
$g = 1$	1/9	1/9	1/9	0	0	0
$g = 2$	1/9	1/9	1/9	0	0	0
$g = 3$	1/9	1/9	1/9	0	0	0
$g = 4$	0	0	0	0	0	0
$g = 5$	0	0	0	0	0	0
$g = 6$	0	0	0	0	0	0

Similarly the $p(g, y|f)$ can be constructed for other values of f 1, 2, 4, 5, and 6. Note that we obtained the entries of this matrix by simply multiplying probabilities $p(g|f)$ and $p(y|f)$ because it is obvious that upon knowing f (hypothetically) the rolls 2 and 3 are independent. And therefore $p(g, y|f) = p(g|f)p(y|f)$ which is the definition of conditional independence. Although the conditional independence is obvious in this example, it will not always be the case with real-world problems considered in this book.

We multiply $p(g,y|f)$'s by $p(f) = 1/6$ and add results obtaining $p(y,g)$:

	$y=1$	$y=2$	$y=3$	$y=4$	$y=5$	$y=6$	
$g=1$	5369	1769	869	469	244	100	
$g=2$	1769	1769	869	469	244	100	
$g=3$	869	869	869	469	244	100	$\times \dfrac{1}{21600}$
$g=4$	469	469	469	469	244	100	
$g=5$	244	244	244	244	244	100	
$g=6$	100	100	100	100	100	100	

Because the above 2D discrete distribution is symmetric upon exchange of rows and columns, it follows that $p(g)$ must be equal to $p(f)$, and can be obtained by summing either the columns or the rows. Once this is done, the values of $p(g,y)$ vs. $p(g)p(f)$ are plotted in Figure 1.8. The plots show that g and y are not unconditionally independent and confirm our intuition about $p(f=1|g=1) > p(f=1)p(g=1)$.

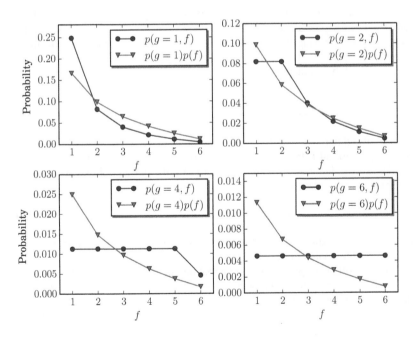

FIGURE 1.8 Comparison of $p(f,g)$ and $p(f)p(g)$ for Example 1.17. Lines connecting the points were added for clarity.

Few of the properties of the conditional independence are specified below. From the definition $g \perp y$ for $p(g, y|f) = p(g|f)p(y|f)$. The alternative definition is

$$p(g|f, y) = p(g|y) \tag{1.22}$$

The above is more telling than Equation (1.21) because it directly indicates that the knowledge of the true value of f is irrelevant for gaining any insights about the true value of g if the true value of y is known. The proof of equivalence of Equation (1.21) and Equation (1.22) is shown below:

$$p(g, y|f) = \frac{p(g, y, f)}{p(f)} = \frac{p(g|f, y)p(f, y)}{p(f)} = p(g|f, y)p(y|f) \tag{1.23}$$

Now since $p(g, y|f) = p(g|f)p(y|f)$ we obtain that $p(g|f, y) = p(g|y)$.

Using similar considerations it can be shown that if $f \perp g|y$ then $g \perp f|y$. An interesting property such that $f, y \perp g_1|g_2$ implies $f \perp g_1|g_2$ and $y \perp g_1|g_2$ can be shown as well (g_1 and g_2 are two different QoIs).

Example 1.18: Three-dice roll (2)–conditional independence

Another example relevant to some properties of imaging systems that will be discussed in Chapter 3 is considered. We have a person A rolling die 1 f and then, as before, rolling die 2 until the number (y) is smaller or equal to the number rolled on die 1. Then, he proceeds to rolling die 3; however, he rolls the die 3 until the number is smaller or the same as the number obtained in roll 2.

$$f \longrightarrow y \longrightarrow g$$

Note the difference with the previous Example 1.17 where roll 3 was done with respect to roll 1. In this example it can be shown that the result of roll 3 is conditionally independent of the result of roll 1 if the number obtained in roll 2 is known. In other words

$$p(g|f, y) = p(g|y) \text{ or } g \perp f|y. \tag{1.24}$$

If y is known the knowledge of f is superfluous and unnecessary and does not bring any additional information about g.

This and other properties of conditional independence listed above can be illustrated using a similar approach as used in Example 1.17. The exercise of demonstrating this is left for the reader.

1.10   SUMMARY

In this chapter the basics of the statistical approach that will be used in this book was introduced. The are two main points that have to be emphasized.

The first is that two stages of knowledge about the problem are identified. The *before experiment* and *after experiment* stages which are denoted as BE and AE. The second is that during the BE stage quantities of interests (QoI) are identified that will be considered and grouped into two categories: (1) quantities that will be observed (OQs) and their true values will be revealed in the experiment and (2) QoIs that will still be unknown in the AE stage and the knowledge about their true values uncertain (UQs). It is assumed that all quantities of interest have a deterministic true value and the knowledge (or uncertainty) about the true values is described by probability distributions.

Once the experiment is performed and probability distributions of known quantities are reduced to delta functions, the probability distributions of UQs (that BE are represented by priors) are modified if there is a statistical dependence between UQs and OQs. The formalism of modification of BE and AE distributions of UQs was introduced by the means of the Bayes Theorem. In order to be able to use the Bayes Theorem it was shown that the model of the experiment (conditional probability of OQs conditioned on UQs) is required as well as the prior probability of UQs. In all of the above, the probability distributions are interpreted as measures of plausibility that a given value of QoI is true [13, 22, 88]. This plausibility can be either a subjective belief which reflects the level of uncertainty in knowledge of the true value of the QoIs, or other distributions chosen for cases where subjective knowledge is poor or there are difficulties in summarizing the knowledge with a distribution.

This lack of objectivity in Bayesian formulation of the statistics is one of the major criticisms of the Bayesian approach. This issue was already discussed in Section 1.7.3 and here we reiterate our view on this subject. In applications of statistical analysis in medial imaging and medicine the ultimate goal of the imaging or medical procedures is to make a decision. In imaging one of the most frequently performed tasks is to summarize imaging data by a single image. This becomes a decision problem because usually there will be many images that could be obtained from data acquired by some physical apparatus that are plausible. For example, we can filter images and by adjusting parameters of the filter providing an infinite number of filtered images at which point we need to decide which of those images should be chosen to accomplish the task at hand. Other decisions are made such as whether disease is present or not, etc.

Any time a decision is made based on uncertain data, the subjectivity must be used to make this decision. In the Bayesian model of uncertain data, the subjectivity is introduced explicitly by the definition of the prior (probability distribution of UQ in BE condition) and by the "loss function" (see Section 2.2) when decisions are made. In classical statistics inferences from the experiment are objective[4]. For example, data may be summarized by the

---

[4]This assuming that the model is correct. In fact, a statistical analysis is always conditional and there are no objective analysis per se, because assumption about the model of the experiment has to be made.

P-value which is the probability of obtaining at least as extreme result as was observed assuming some hypothesis (or model of the experiment) is true. However, in order to make a decision (reject the hypothesis or not) the *significance* (the value of the threshold) has to be selected BE. This selection is a subjective choice that should be varied based on likelihood of the hypothesis.

The misunderstanding of the pseudo-objectivity of classical statistics leads to many incorrect conclusions found in medical literature [35, 46, 89, 101]. For example, the classical hypothesis testing used extensively in medical imaging and medicine, the experimental evidence that is summarized by the classical methods by the P-value should always be evaluated in the light of plausibility of the hypothesis (see the next example); however, the classical statistics provide only limited means for quantitative combination of the findings from the data and plausibility of the hypothesis, and plausibility of hypothesis is seldom discussed in the context of objective evidence summarized by the P-value.

Example 1.19: Summary of experimental evidence: P-values

Suppose we roll a six-sided die thee times and we obtain six all three times. Using measure of classical statistics the null hypothesis that obtaining any number is equally likely can be rejected with a high statistical significance (low P-value). Therefore the objective analysis of the data says that the die is rigged. Obviously any reasonable decision maker who uses prior knowledge would require much substantial evidence to be convinced that the die is "unfair" and based on this subjective judgment, the null hypothesis will not be rejected based on the mentioned experiment. In order to do so, the rejection region, has to be equal to much lower value than P=0.05 or P=0.01 used so extensively in the field. Perhaps for the example with die the significance level should be set at 10^{-6} level. Looking through scientific literature, it is rare that investigators ever discuss the reasoning behind choosing the classical statistical significance level used in their work.

We argue that the use of Bayesian methods (e.g., as the BEAE view) is justified and subjectivity is unavoidable when making decisions about uncertain events. The use of Bayesian approaches force investigators to express their beliefs in a form of the prior rather than combine in some unspecified way the experiential evidence (summarized for example with the P-value) with prior beliefs to make decisions. In medical imaging, decision making is the end-point of any imaging procedure (e.g., disease present or not, disease progresses or not, etc.) and therefore the issue of combining the experimental evidence with the prior beliefs in some coherent way is of utmost importance.

2 Elements of decision theory

2.1 INTRODUCTION

The ultimate goal of medical imaging is to answer a clinically relevant questions about the disease status of patients. As an example, one of the frequent tasks is to determine if a patient who undergoes medical imaging procedures has cancer or not. For this decision problem there are two possible outcomes: cancer present or not. Another common task is to determine the concentration of a tracer in the organ of interest. For this case there is usually an infinite number of outcomes as the concentration is a continuous variable. Another task may be to determine if the concentration of tracer changes in longitudinal studies. The change can be classified as substantial, marginal, or no change in which case three types of outcomes are identified. Since the data acquired by nuclear imaging medical scanner is inherently random, a clear choice of the optimal decision will seldom be available. The goal of decision theory is to provide tools that can be used for making "the best" decision based on available relevant information.

We differentiate the decision making into two categories depending on the level of knowledge available. The more certain (more information is available) regime of AE stage for decision specific to a given data set is described in Section 2.3, and decisions that involve analysis in BE stage where data is not yet revealed, in Section 2.4.

Based on the BE/AE concept, we identify two states of knowledge that are available for decision making. In the BE stage, the data (the result of the experiment) is not known. Since it is expected that upon seeing the data (likelihood function is determined) the uncertainty about unobservable quantities (UQs) will be reduced and observable quantities (OQs) will be known, it is rational to choose to make decisions in AE condition. The AE is our main decision-making stage. Sometimes, however, we want also to make decisions that are not associated with any particular data set. Grading imaging system performance would be an example of such task. Relying just on a single experiment (single data set) does not seem to be the best approach because the conclusions would be dependent on the data (OQs), which in general would not guarantee that the same conclusions are reached when different data are obtained. It is therefore desirable to use BE conditions for this task and consider all possible data sets that can be obtained by the system.

We will quite generally describe a decision using the mathematical symbol δ. For example, in a binary decision (the choice between two alternatives) about cancer present or not the δ have two values which indicate decisions made in favor of cancer present or in favor that it is not. We can adopt a convention that $\delta = 1$ indicates that the cancer is present and $\delta = 0$ that it is not present. This can be extended to *multiple alternatives decision making* and for example the third and fourth categories can be introduced corresponding probably yes and probably not decisions. In such case the $\delta \in \{0, 1, 2, 3\}$ where $\{0, 1, 2, 3\}$ are arbitrarily assigned to some possible decision. For example, the following assignment can be used: 0–no, 1–yes, 2–probably yes, 3–probably no. The numerical values are assigned but they do not carry any quantitative information and should be considered as simple *labels* assigned to a certain decisions. In the estimation decision problem, the δ represents the decision about the best values of some unobservable quantities.

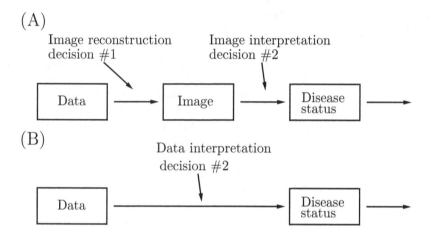

FIGURE 2.1 (A) Simplified model of the traditional two-step decision-making process in interpretation of medical nuclear imaging data. (B) An alternative approach in which the formation of images is not used and decisions are made based directly on the data.

In medical imaging we will frequently be faced with the task to represent the data in the form of an image that can be further processed in a decision-making chain (Figure 2.1). In nuclear imaging the image is formed that represents concentrations of radioactive tracer in voxels. The formation of the image is a decision problem because of the uncertainty in available information (decision #1 in Figure 2.1(A)). This image is interpreted by a physician (decision #2 is made) who makes a decision about disease status. Unfortunately, the physician who makes this decision is frequently not the final decision maker about the course of therapy or at least is not the sole final decision maker, and therefore more decisions are made and more sources of uncertainty play a role in the final outcome. The issue of decision making in medicine is complex and beyond the scope of this work but the interested reader is encouraged to consult Hunink [45] and Mushlin and Greene II [75]. We will be concerned in this book only with decision #1 and decision #2 as shown in Figure 2.1(A).

In the traditional decision chain described in Figure 2.1(A), the optimality of such approach is achieved by optimization of the image reconstruction step in order to achieve optimal results in terms of assessment of disease status. There is a substantial literature on this topic and the approach is referred to as the task-oriented image assessment [6]. This approach is difficult to implement in practice because decision #2 is based on the result of decision #1. This two-step process creates severe computational difficulties in representation of data uncertainties in reconstructed images (intermediate step). In this book we will not follow this approach.

In our approach, the goal is to eliminate the two-step decision process as shown in Figure 2.1(A) and use just a single decision based directly on the data as shown in Figure 2.1(B).

2.2 LOSS FUNCTION AND EXPECTED LOSS

This section introduces the new concept of the *loss function* which is required for the implementation of the decision theory. The loss function quantifies the consequences of the decision-making process. If a procedure uses the loss function it will be referred to as a *decision-theoretic* procedure. The significance of the loss function is illustrated with the following Example 2.1:

Example 2.1: Balls in the urn

Suppose a decision maker plays a game of chance using balls in the urn as illustrated in Figure 2.2 and draws exactly one ball. Before the game (BE stage) he has to bet (make a decision based on uncertain knowledge) on a number that will be drawn. The total number of balls and how many balls have given numbers are known to the decision maker. In this example there are a total of three balls. Two have number 1 and one ball has number 2. Interestingly, if no other rules of the game are specified we should conclude that the decision maker does not have enough information to make a rational decision (bet, guess) about the number that will be drawn from the decision theory point of view. The missing piece of information is the consequence of the decision. Only when the loss function is known, a rational decision can be made. The consequences describe a gain/loss when the decision is right or wrong.

As illustrated in the above example, a decision problem (deciding which ball will be drawn) that involves uncertainty (it is unknown which ball will be drawn) cannot be solved without specification of the consequences of the decision (the loss function[1]). In Example 2.1 it is tempting to use the *heuristic*

---

[1]The loss function can also be negative indicating gains.

FIGURE 2.2 The urn with three balls. The experiment consists of random drawing of one of the balls.

[55]. The heuristic is an experience-based approach to decision making which results in a decision that is not guaranteed to be optimal. Using the heuristic without the explicit definition of the loss function we may decide in favor of ball 1 since the odds of drawing the ball with number 1 are higher than drawing the ball with number 2. This, however, would be an error and not necessarily an optimal from the decision theory point of view. To illustrate this point, let's consider the continuation of the previous example:

(A) decision

true	$\delta = $ Yes	$\delta = $ No
1	-100	50
2	100	-100

(B) decision

true	$\delta = $ Yes	$\delta = $ No
1	-100	0
2	100	-100

FIGURE 2.3 Two definitions of the loss function shown in (A) and (B). The "true" corresponds to the result of the draw, and $\delta = $ Yes corresponds to the decision that the ball with number 1 was drawn.

Example 2.2: Balls in the urn (2)

Example 2.1 is continued by adding of the loss function. Suppose that we specify the loss function as illustrated in Figure 2.3.

The two tables in Figure 2.3 show two different loss functions specified for the decision problem described in Example 2.1. The decision $\delta = $ Yes corresponds to deciding about drawing a ball with 1 and the decision $\delta = $ No corresponds to predicting that a ball with 2 is drawn. Which of these two decisions is correct from the decision theory point of view considering the loss is incurred as shown in the tables?

It is counter-intuitive that the negative loss is something desirable. However, it can be interpreted as something opposite to penalty which makes the negative sign more perceptive. The negative loss can be considered as the reward and the positive loss quantifies the penalty that is sustained. For the decision problem described in this example, there are four possibilities (four cells in the tables) of the final outcome of the decisions.

'Yes' and 1 We can decide 'Yes' (a ball with 1 will be drawn) and in fact a ball with 1 is drawn (upper left cell in Figure 2.3(A)). If this happens we are rewarded with negative loss -100.

'Yes' and 2 If we decide 'Yes' and then a ball with 2 is drawn the loss incurred is 100 (lower left cell in Figure 2.3(A)). We made an incorrect decision and in the rational decision theory we are penalized.

'No' and 2 We can decide 'No' (other than a ball with 1 is drawn, which in this case means that a ball with 2 is drawn) and in fact a ball with 2 is drawn (lower left cell in Figure 2.3(A)). If this happens we are rewarded with the negative loss -100, which is the same as the reward for being right in 'Yes' and 1 case.

'No' and 1 If we decide 'No' and then a ball with 1 is drawn anyway the loss incurred is 50 (upper right cell in Figure 2.3(A)).

Since drawing of a ball with 1 is twice more probable as drawing a ball with 2 it is quite straightforward to select the optimal decision. We define the *optimal decision* as the one with the minimal expected loss. For the loss function shown in Figure 2.3(A) the expected loss of the decision 'Yes' is $2/3 \times -100 + 1/3 \times 100$ because drawing #1 is two times more probable than drawing #2. It follows that the *expected loss* of decision 'Yes' is equal to -33. The expected loss of decision 'No' is 0 and therefore the optimal decision is 'Yes'

For the loss specified in Figure 2.3(B) for both decisions, 'Yes' and 'No', the expected loss is equal to 33 and therefore both decisions are equivalent and both are optimal from the point of view of the decision theory. Interestingly, although both decisions are equivalent, a human decision maker would strongly prefer decision 'No' because of the loss-aversion heuristics. By selecting 'No' the decision maker would assure that there is no loss (no positive value of the loss function) which case is strongly preferred option for the human decision maker. In the case of decision 'Yes', positive loss is possible and it would be a less preferable option for the human decision maker. The loss-aversion heuristics is a part of the *prospect theory* developed by Kahneman and Tversky [54].

Example 2.2 illustrates the concept of the optimal decision making. We will put it now in a more formal mathematical context. Let the vector φ define all quantities of interest (QoI) which are uncertain at the time of decision making. Before experiment (BE) this would be all observable quantities (OQs) and unobservable quantities (UQs). After the experiment (AE) only the UQs are uncertain and therefore they only would be represented in φ. The φ is a vector and the space of all possible values of φ is $\boldsymbol{\Psi}$.

We define the *loss function* as $L(\varphi, \delta)$, which for each possible value of uncertain quantities φ assigns a penalty assuming the decision is δ. The loss function is a real-valued scalar. Suppose that at the time of decision making the probability that φ is true is known and equal to $p^*(\varphi)$. From the definition of probability we have that $\int_{\varphi \in \Psi} p^*(\varphi) = 1$. We define the *expected loss function* for the decision δ as $\varrho(\delta)$ as

$$\varrho(\delta) = \int_{\varphi \in \Psi} L(\varphi, \delta) p^*(\varphi) \tag{2.1}$$

Optimal decision principle: The optimal ("best") decision from all possible decisions $\delta \in \mathcal{D}$ is such that it minimizes expected loss ϱ. By $\mathcal{D}$ we denote all possible decisions.

Example 2.3: Balls in the urn (3)

Continuing Example 2.3 φ represents the result of the draw and therefore it is a scalar that can have only two values from the set $\{1, 2\}$. The decision problem in the example can only be considered at the BE stage because after the experiment (the draw) there are no uncertain quantities and the decision making is trivial is such case. We see the $\varphi \in \{1, 2\}$ and the possible decisions are $\mathcal{D} = \{'Yes', 'No'\}$. It follows that for the example in Figure 2.3(A) the loss function $L(1, 'Yes') = -100$, $L(1, 'No') = 50$, $L(2, 'Yes') = -100$, and $L(2, 'No') = -100$. The $\varrho('Yes') = -33$ (computed using Equation (2.1)) and $\varrho('No') = 0$ and therefore the $\mathcal{D} = 'Yes'$ is the optimal decision.

2.3   AFTER-THE-EXPERIMENT DECISION MAKING

In the after-the-experiment decision making, we will use the posterior to compute the posterior expected loss using Equation (2.1). If the observed QoIs are denoted by $\mathbf{g}$, then the $p^*(\varphi)$ as defined earlier in this chapter will be equal to the posterior of UQs $p(\boldsymbol{\theta}|\mathbf{G} = \mathbf{g})$ where $\boldsymbol{\theta}$ represents all unobservable quantities. It follows that decisions will be based on the *posterior expected loss* defined as

$$\varrho_{\text{post}}(\delta; \mathbf{G} = \mathbf{g}) = \int_{\boldsymbol{\theta} \in \Theta} L(\boldsymbol{\theta}, \delta) p(\boldsymbol{\theta}|\mathbf{G} = \mathbf{g}) \tag{2.2}$$

We note that the posterior expected loss is a function of observed QoI $\mathbf{g}$.

Example 2.4: Conditional two-dice roll (5)

This is the continuation of Example 1.10. For convenience the description is repeated here and the symbol of UQ f used before is replaced here with the symbol θ to be consistent with the notation used in this chapter.

Person A rolls the first die obtaining result θ and then rolls the second die obtaining result g until the number is equal or less than the result of the first roll.

TABLE 2.1

Values of the pre-posterior of θ conditioned on g where g is assumed to be observed in the experiment (dice rolls)

$p(\theta|g)$

	$\theta \to$	1	2	3	4	5	6
$g \downarrow$	1	60/147	30/147	20/147	15/147	12/147	10/147
	2	0	30/87	20/87	15/87	12/87	10/87
	3	0	0	20/57	15/57	12/57	10/57
	4	0	0	0	15/37	12/37	10/37
	5	0	0	0	0	12/22	10/22
	6	0	0	0	0	0	1

Only the final number obtained on the second die g is revealed to the observer. The pre-posterior $p(\theta|g)$ is determined in Example 1.13 and is repeated here in Table 2.1.

The task is to make a decision about what is the number on the first roll ($\delta \in \{1, 2, 3, 4, 5, 6\}$) under the quadratic loss function $L(\theta, \delta) = (\theta - \delta)^2$. To make it more intuitive, suppose it is a game and the rules of the game can be explained such that if we make a decision about θ and we are mistaken we will have to to pay $(\theta - \delta)^2$ dollars. In the best case scenario, we come even (no penalty or reward) and if we fail to make the correct guess we pay. For example, if we guess that θ is 3 where it is actually 1 we will have to pay $(3 - 1)^2 = 4$ so we will be penalized with the $4 penalty.

As explained in this chapter the optimal decision is such that the expected posterior loss is minimized. Since the decision depends on the observed data in general it will be different for each observed value of g. In Table 2.2 the posterior expected loss is computed and optimal decision about the θ is selected for every possible data g that can be measured. Therefore, the table actually contains the description of six independent decision problems each corresponding to a row in the table. The computation follows Equation (2.2).

2.3.1 POINT ESTIMATION

One of the most common tasks performed in medical imaging is the estimation of the UQs. Since the UQs are uncertain, the estimation is a process of making a decision about the true value of UQs. In the Bayesian paradigm introduced in Section 2.2, the optimal decisions (that includes estimations) are obtained from minimizing the posterior loss and therefore the loss function must be specified in order to define the estimator. Specification of the loss function is not an easy task in many situations and must be considered on a case-by-

TABLE 2.2

Values of the posterior expected loss for the decision δ corresponding to pre-posterior in Table 2.1 and the quadratic loss

	$\delta \rightarrow$	$\varrho(\delta; g)$						Optimal δ
		1	**2**	**3**	**4**	**5**	**6**	
$g \downarrow$	**1**	4.67^a	2.76	2.88	4.98	9.08	15.18	2
	2	7.90	4.00	2.10	2.21	4.31	8.41	3
	3	11.52	6.11	2.68	1.26	1.84	4.42	4
	4	15.59	8.86	4.14	1.41	0.68	1.96	5
	5	20.09	12.18	6.27	2.36	0.45	0.54	5
	6	25.00	16.00	9.00	4.00	1.00	0.00	6

^{a}Example calculation of the posterior expected loss for $g = 1$ and $\delta = 1$:
$4.67 = (60/147) \times 0 + (30/147) \times 1 + (20/147) \times 4 + (15/147) \times 9 + (12/147) \times 16 + (10/147) \times 25$

case basis depending on a task at hand. When choosing a loss function, the complexity has to be weighted against the computational requirements.

From the point of view of computational requirements, there are two loss functions that are implementable in most problems and in fact are used extensively in the field of Bayesian estimation. The most popular loss function used by Bayesians is the *quadratic loss* which we have already used in Example 2.2. The quadratic loss is defined as:

$$L(\boldsymbol{\theta}, \boldsymbol{\delta}) = (\boldsymbol{\theta} - \hat{\boldsymbol{\delta}})^T (\boldsymbol{\theta} - \hat{\boldsymbol{\delta}}) \tag{2.3}$$

where $\hat{\boldsymbol{\delta}}$ indicates the point estimator of $\boldsymbol{\theta}$. Since QoIs are numbers, this loss function corresponds to the Cartesian distance between vectors $\boldsymbol{\theta}$, and $\hat{\boldsymbol{\delta}}$ so the larger the distance between the estimate of the true value, the higher the value of the loss function. Therefore, this loss function reflects the belief that the amount of error that is made when choosing the estimate matters. In other words a smaller error is better than a bigger error. Although this may seem trivial this reasonable property of the loss function is not used for the other very popular Bayesian estimator discussed later in this section for which the amount of error is irrelevant.

The point estimators $\hat{\boldsymbol{\delta}}$ have continuous values. Even if the QoI $\boldsymbol{\theta}$ has discrete values, the point estimators are continuous. To estimate the discrete value of $\boldsymbol{\theta}$, a different decision problem of hypothesis testing must be used and is described in Section 2.3.3.

In order to find the mathematical form of the estimator that minimizes the posterior expected loss with the quadratic loss function, we differentiate the posterior expected loss with respect to $\hat{\boldsymbol{\delta}}$ and find the root by

$$\frac{d}{d\hat{\boldsymbol{\delta}}} \int_{\boldsymbol{\theta} \in \Theta} (\boldsymbol{\theta} - \hat{\boldsymbol{\delta}})^T (\boldsymbol{\theta} - \hat{\boldsymbol{\delta}}) p(\boldsymbol{\theta}|\mathbf{G} = \mathbf{g}) = 0. \qquad (2.4)$$

The solution of the above is the posterior mean $\boldsymbol{\mu}(\mathbf{G} = \mathbf{g})$

$$\boldsymbol{\mu}(\mathbf{G} = \mathbf{g}) = \int_{\boldsymbol{\theta} \in \Theta} \boldsymbol{\theta} p(\boldsymbol{\theta}|\mathbf{G} = \mathbf{g}) \qquad (2.5)$$

The posterior mean is the *minimum mean square error* (MMSE) estimator and is the most frequently used Bayesian estimator. This popularity comes from the fact that computation of the expectation over some distribution is relatively straightforward (at least conceptually) and because the quadratic loss functions make sense for many applications. In order to provide the error of the estimation, which indicates the accuracy of the estimate, the *posterior covariance* $\mathbf{V}(\mathbf{G} = \mathbf{g})$ is used and defined as

$$\mathbf{V}(\mathbf{G} = \mathbf{g}) = \int_{\boldsymbol{\theta} \in \Theta} (\boldsymbol{\theta} - \hat{\boldsymbol{\delta}})(\boldsymbol{\theta} - \hat{\boldsymbol{\delta}})^T p(\boldsymbol{\theta}|\mathbf{G} = \mathbf{g}). \qquad (2.6)$$

The posterior covariance is the most general estimate of error of the Bayesian point estimates. Note that posterior standard error of the component of θ_i is $\sqrt{V_{ii}}$ [8].

TABLE 2.3

MMSE estimators with the posterior standard errors

be-3pt]		MMSE estimator of θ	Posterior variance	Error of the MMSE estimator
$g \downarrow$	1	2.45[a]	2.57	1.61
	2	3.45	1.90	1.38
	3	4.21	1.22	1.11
	4	4.86	0.66	0.81
	5	5.45	0.25	0.50
	6	6.00	0.00	0.00

[a] Example calculation of the posterior mean (MMSE estimator) for $g = 1$:
$2.45 = (60/147) \times 1 + (30/147) \times 2 + (20/147) \times 3 + (15/147) \times 4 + (12/147) \times 5 + (10/147) \times 6$

Example 2.5: Conditional two-dice roll (6)

Based on the pre-posterior shown in Table 2.1 (each row can be interpreted as the posterior) the estimator for each value of data g is calculated. The posterior covariance and posterior standard errors are also computed.

When MMSE estimators are rounded to the nearest integer and compared to optimal decisions made in Example 2.2, a perfect match is obtained. Ideally for discrete QoI one would like to compute the posterior expected loss for every possible θ (as was done in Example 2.2) and then decide based on the lowest value of the posterior expected loss. This approach however (for most of realistic applications) will not be practical because of θ will typically be a high-dimensional space and the task of computing posterior expected loss for every possible value from Θ will be impossible. In such case use of the MMSE estimator is an attractive option.

Another Bayesian estimator used frequently in imaging is called *maximum a posteriori* (MAP) estimator. The loss function that leads to MAP estimator has to be defined differently for discrete and continuous θ. For the continuous θ the loss function leading to MAP estimator is

$$L(\boldsymbol{\theta}, \boldsymbol{\delta}) = \begin{cases} 0 & \text{if } ||\boldsymbol{\theta} - \boldsymbol{\delta}|| \leq \epsilon \\ 1 & \text{if } ||\boldsymbol{\theta} - \boldsymbol{\delta}|| > \epsilon \end{cases} \tag{2.7}$$

where ϵ is some small constant. For the discrete case, the so-called $0 - 1$ *loss* leads to the MAP estimator and it is defined as

$$L(\boldsymbol{\theta}, \boldsymbol{\delta}) = \begin{cases} 0 & \text{if } \boldsymbol{\theta} = \boldsymbol{\delta} \\ 1 & \text{if } \boldsymbol{\theta} \neq \boldsymbol{\delta} \end{cases} \tag{2.8}$$

It is straightforward to show that for $\epsilon \to 0$ the MAP estimator corresponds to the node[2] of the posterior (estimator that minimizes the posterior expected loss) for continuous and discrete cases. Although the MAP estimator is very popular we have to keep in mind the limitations of this estimator. The first obvious limitation is non-uniqueness. If the posterior has many nodes then there are as many equivalent MAP estimators. Another issue with the MAP estimator that we have to keep in mind and which may be important for some applications is its peculiar shape of the loss functions given by Equations (2.7) and (2.8) and presented in Figure 2.4. No matter how much we are mistaken in estimating the true value of θ, the same value of the loss function is incurred. For example, in Figure 2.4(B) the same loss is incurred whether estimator $\hat{\theta}_1$ (close to the true value) and estimator $\hat{\theta}_2$ (far from the true value) are considered. This property of the loss function leading to the MAP estimator has to be always kept in mind and the appropriateness of the estimator considered when approaching a problem. For some applications such loss would actually be correct but for the majority of problems it does not correspond to our actual perception of loss in a given problem. To illustrate this let's consider

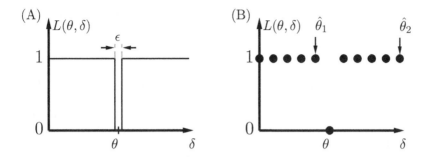

FIGURE 2.4   The one-dimensional loss function that leads to MAP estimators for continuous (A) and discrete (B) cases.

the following two examples:

Example 2.6: Pulling goaltender in an ice hockey game

Suppose a hockey team A is down 1 goal losing to team B and there are 30 seconds left in a play off game (losing team is out of competition). Suppose we consider only three outcomes: (1) Team A scores keeping its changes alive and forcing the overtime, (2) Team B scores winning the match by the difference of two goals, and (3) there are no more goals and team B wins by one goal. Therefore $\theta = 1, 2, 3$. The coach of team A has the decision to make whether to substitute the goaltender for an extra attacker. $\delta = 1$ corresponds to the decision of substitution and $\delta = 0$ of keeping the goaltender. Suppose that the hypothetical probability that either team scores without substitution is 1%. If the substitution is made, the probability that team A scores is increased to 1.2% and that team B scores is increased to 10%.

Since the chance of scoring for the opposite team increase so much it may seem that decision to substitute is not a good idea. Let's see if we can explain with decision theory the fact that the decision to substitute is frequently made by coaches in such a situation (assuming that they follow decision-theoretic approach). Formally, the loss function is $L(\theta = 1, \delta = 1) = 0$ which corresponds to a case that the coach opts for the substitution and team A scores. In such case there is no loss. For the other two cases, the loss function is the same $L(\theta = 2, \delta = 1) = L(\theta = 3, \delta = 1) = x_1$. Similarly, $L(\theta = 1, \delta = 0) = 0$ and $L(\theta = 2, \delta = 0) = L(\theta = 3, \delta = 0) = x_0$.

Based on those assumed values of the loss function, the posterior expected losses are $\varrho(1) = 0.988x_1$ and $\varrho(0) = 0.99x_0$ where x_1 and x_0 are some constants that represent some losses of prestige for the coach or loss of his

---

[2]The node is the maximum of function and in this case maximum of the distribution.

job, etc. Therefore, the substitution is the optimal decision because it has the smaller posterior expected loss if we assume $x_1 = x_0$.

Assumed condition that $x_1 = x_0$ may not be true in reality. For example, the coach should consider that since he is expected to pull the goaltender in a situation like the one described in this example, then if he does not do that and team A loses, he will be more strongly criticized than in the case when he pulls the goalkeeper and team A loses. In such case $x_1 < x_0$. This condition only reinforces the decision of pulling out the goaltender. Readers may speculate other possibilities for the loss function. For example, there may be different reward (negative loss $L(1, 0) < 0$ and $L(1, 1) < 0$) for scoring with and without the substitution, etc.

Example 2.7: Conditional two-dice roll (7)

In general, the MAP estimators are much different than the results obtained by the MMSE estimators when the posterior probability distributions are not symmetrical around the mean.

The posteriors (rows in Table 2.1) are asymmetrical and because of that the estimators (MMSE and MAP) differ considerably. When the posterior is approximately symmetrical both estimators are similar. In particular for normal distributions which are symmetric the MAP is equal to MMSE.

TABLE 2.4

Comparison of the lowest posterior expected loss, MMSE, and MAP estimators

		Lowest expected loss	MMSE	MAP
$g\downarrow$	1	2	2.45	1
	2	3	3.45	2
	3	4	4.21	3
	4	5	4.86	4
	5	5	5.45	5
	6	6	6.00	6

2.3.2 INTERVAL ESTIMATION

An alternative to the point estimation is the *interval estimation*. Rather than providing a single solution that is the "best" guess of what are the true values of uncertain quantities, the interval estimation provides a range of values that contain the true value with a high probability. We will refer to such region as *credible set* or *credible interval*. We define the credible set following Berger [8] as

Definition: A x% credible set for $\boldsymbol{\theta}$ is a subset $\boldsymbol{\Theta}^x$ of $\boldsymbol{\Theta}$ such that

$$x \leq p(\boldsymbol{\Theta}^x|\mathbf{G} = \mathbf{g}) = \int_{\boldsymbol{\theta} \in \boldsymbol{\Theta}^x} p(\theta|\mathbf{G} = \mathbf{g}) \tag{2.9}$$

There are two important observations that we make about the definition of the credible set. First, the interval definition does not use a loss function and therefore it is not meaningful from the decision-theoretic perspective. The credible set should be interpreted as a convenient summary of the posterior but it cannot be used directly for decision making. The other observation is that the credible set is not unique and there will be a large (if not infinite) number of credible sets for a given posterior. A popular approach to reduce the non-uniqueness is to include in the set only such θ's which have a high value of probability. This leads to the *highest posterior density* (HPD) confidence set defined as in Berger [8]:

Definition: A x% HPD credible set for $\boldsymbol{\theta}$ is the subset $\boldsymbol{\Theta}^x$ of $\boldsymbol{\Theta}$ such that

$$\boldsymbol{\Theta}^x = \boldsymbol{\theta} \in \boldsymbol{\Theta} : p(\boldsymbol{\theta}|\mathbf{G} = \mathbf{g}) \geq \varepsilon(x), \tag{2.10}$$

where $\varepsilon(x)$ is the largest constant such that

$$p(\boldsymbol{\Theta}^x|\mathbf{G} = \mathbf{g}) \geq x. \tag{2.11}$$

The credible sets are quite easy to compute especially for unimodal posteriors (see Example 2.8). In general, however, the credible sets do not have to be continuous and may have disjoint regions which may look unusual, but it would typically indicate a clash between the prior and the data. It is important to recognize the situation when this happens and disjointed credible sets alert us about this possibility.

Example 2.8: Credible sets

Suppose the posterior has the following gamma distribution form $p(\theta) = \theta e^{-\theta}$ defined on $\boldsymbol{\Theta} : \theta \in [0, \infty]$. We have to specify the 95% HDP confidence set. This is asymmetric but unimodal distribution and therefore the HDP confidence set will not be disjoint. We define $\boldsymbol{\Theta}^{95\%}$ as range $[\theta_1, \theta_2]$ such that

$$\int_{\theta_1}^{\theta_2} \theta e^{-\theta} d\theta = 0.95. \tag{2.12}$$

An useful approach to find the HDP set with unimodal posteriors is to look for such θ_1 and θ_2 for which the $\theta_2 - \theta_1$ is at the minimum, but the condition expressed by Equation (2.12) holds. For unimodal posterior, this definition of HDP is equivalent to the general definition given earlier in this section. It is also easier to compute.

2.3.3 MULTIPLE-ALTERNATIVE DECISIONS

The case of *multiple-alternative decisions* applies to situations where a single choice needs to be selected from a group of alternatives. There is a little difference in Bayesian computing between binary- and multiple-alternative decision making. The optimal decision is simply such that it has the smallest posterior expected loss. In Example 2.4 we have already used multiple-alternative decisions to decide about the number on the first roll of the die.

The simplest and a common special case of a decision task is the binary decision where there are two choices available and the task is to choose one of them. A *detection problem* in which the signal-present or signal-absent decision has to be made is an example of binary decision problem. In nuclear imaging, the detection of cancer or other diseases using imaging data is an example of a binary decision. Because the binary decision making is so important, we discuss it in detail in Section 2.3.4.

To put it into a more formal form we assume that at the time of decision making, quantities $\boldsymbol{\theta} \in \boldsymbol{\Theta}$ are uncertain. We also specify the N alternatives of decisions to be made where $N \geq 2$ such that $H_n : \boldsymbol{\theta} \in \boldsymbol{\Theta}_n{}^3$ where $\boldsymbol{\Theta}_n$ is the subset of $\boldsymbol{\Theta} = \bigcup_{n=1}^{N} \boldsymbol{\Theta}_n$ and $\boldsymbol{\Theta}_n \cap \boldsymbol{\Theta}_{n'} = \emptyset$ for $n \neq n'$.

We define the loss function of multiple-alternative decisions as

$$L(\boldsymbol{\theta}, \delta(H_n)) = \sum_{n'=1}^{N} L_{n'n} \chi_{\boldsymbol{\Theta}_{n'}}(\boldsymbol{\theta}) \qquad (2.13)$$

where by $\delta(H_n)$ we indicate the decision in favor of hypothesis (choice) n. The value of $L_{n'n}$ is the value of the loss incurred when making a decision in favor of hypothesis n when n' is the correct decision. The *membership function* χ is defined as

$$\chi_{\boldsymbol{\Theta}_n}(\boldsymbol{\theta}) = \begin{cases} 1 & \text{for } \boldsymbol{\theta} \in \boldsymbol{\Theta}_n \\ 0 & \text{for } \boldsymbol{\theta} \notin \boldsymbol{\Theta}_n \end{cases} \qquad (2.14)$$

Using the loss function defined by Equation (2.13) the posterior expected loss of decision $\delta(H_n)$ in favor of the hypothesis n is defined as

$$\varrho(\delta(H_n)) = \int_{\boldsymbol{\theta} \in \boldsymbol{\Theta}} L(\boldsymbol{\theta}, \delta(H_n)) p^*(\boldsymbol{\theta}) \qquad (2.15)$$

---

[3]We use symbol H (the first letter in *hypothesis*) to indicate one choice among all possible alternatives. It is therefore a hypothesis that a given alternative is true.

where the $p^*(\boldsymbol{\theta})$ is the probability of quantity $\boldsymbol{\theta}$ at time of decision making. Using Equation (2.13) we have

$$\varrho(\delta(H_n)) = \sum_{n'=1}^{N} L_{n'n} \int_{\boldsymbol{\theta} \in \boldsymbol{\Theta}} \chi_{\boldsymbol{\Theta}_{n'}}(\boldsymbol{\theta}) p^*(\boldsymbol{\theta})$$

$$= \sum_{n'=1}^{N} L_{n'n} \int_{\boldsymbol{\theta} \in \boldsymbol{\Theta}_{n'}} p^*(\boldsymbol{\theta}) = \sum_{n'=1}^{N} L_{n'n} p^*(\boldsymbol{\Theta}_{n'})$$

where we used notation $p^*(\boldsymbol{\Theta}_{n'}) = \int_{\boldsymbol{\theta} \in \boldsymbol{\Theta}_{n'}} p^*(\boldsymbol{\theta})$ indicating the probability that the true value of uncertain quantity $\boldsymbol{\theta}$ is in the set $\boldsymbol{\Theta}_{n'}$. Since we are concerned in this section with the AE decision making, the posterior is used to make decisions and therefore the posterior expected loss of decision in favor of the hypothesis n (the $p^*(\boldsymbol{\Theta}_{n'})$ is replaced with $p^*(\boldsymbol{\Theta}_{n'} | \mathbf{G} = \mathbf{g})$) is

$$\varrho(\delta(H_n); \mathbf{G} = \mathbf{g}) = \sum_{n'=1}^{N} L_{n'n} p(\boldsymbol{\Theta}_{n'} | \mathbf{G} = \mathbf{g}). \qquad (2.16)$$

Multiple-alternative Bayesian decision making: In Bayesian computing the decision in favor of hypothesis/choice H_n is made when the posterior expected loss for H_n is the smallest among all alternatives.

Example 2.9: Conditional two-dice roll (8)

Let's examine Example 2.4 from the point of view of multiple-alternative decision making. In this example the multiple-alternative was defined as each of the six alternatives/hypotheses corresponded to a different number obtained in roll one. It is therefore an example of *simple hypothesis testing* where there is one–to–one correspondence between the number of unknown quantities and the number of alternatives/hypotheses. In this example there were six possible numbers (values of θ) and therefore six hypotheses.

Suppose, however, that we have only two hypotheses. Hypothesis H_1 is that the number is either 1, 2, 3, or 4, and hypothesis H_2 that the number is 5 or 6. The posteriors of those hypotheses can quickly be calculated from Table 2.1 by summing probabilities in columns corresponding to the hypotheses. The loss function has to be specified as well and we just assume simple loss function of being zero when we are right and equal to x when we are wrong. Therefore, $L_{11} = L_{22} = 0$ and $L_{12} = L_{21} = x$. With those assumptions the calculation of the posterior expected losses for different values of observed numbers on the second roll are given in Table 2.5. The hypothesis when there are many different values of quantities of interest (QoI) that correspond to the same hypothesis is called *composite hypothesis*. In our case hypothesis 1 corresponds to four possible values of uncertain quantities and hypothesis 2 corresponds to two values. Table 2.5 shows the result of calculation of the posterior expected losses for both hypotheses. Based on the posterior expected loss (last two columns in Table 2.5) the decision in favor of hypothesis 1 should be made when the second roll g is 1, 2, or 3, and in favor of hypothesis 2 when the second roll is 4, 5, or 6.

TABLE 2.5

Values of posterior expected losses for composite hypotheses

		$p(\Theta_1|G=g)$	$p(\Theta_2|G=g)$	$\varrho(\delta(H_1);G=g)$	$\varrho(\delta(H_2);G=g)$
$g\downarrow$	1	$125/147$	$22/147$	$22x/147$	$125x/147$
	2	$65/87$	$22/87$	$22x/87$	$65x/87$
	3	$35/57$	$22/57$	$22x/57$	$35x/57$
	4	$15/37$	$22/37$	$22x/37$	$15x/37$
	5	0	1	x	0
	6	0	1	x	0

2.3.4　BINARY HYPOTHESIS TESTING/DETECTION

The previous section discussed the general multiple-alternative decision making that includes binary *hypothesis testing*. We dedicate here a separate section to discussion of the binary hypothesis testing to emphasize its importance as well as to derive correspondence with well-known approaches used in the imaging field which are special cases of the general Bayesian decision theory. Because only two hypotheses are considered, in order to simplify and improve clarity of the notation we define the following $p_1 = p(\Theta_1|G = g)$ and $p_2 = p(\Theta_2|G = g) = 1 - p_1$ as the posterior probabilities of hypotheses 1 and 2. The posterior expected losses for decisions 1 and 2 are denoted as $\varrho_1 = \varrho(\delta(H_1))$ and $\varrho_2 = \varrho(\delta(H_2))$. Using the new notation we have

$$\varrho_1 = L_{11}p_1 + L_{21}p_2 \qquad (2.17)$$

and

$$\varrho_2 = L_{12}p_1 + L_{22}p_2 \qquad (2.18)$$

The optimal decision maximizes the posterior expected loss and thus the decision in favor of hypothesis 1 is made when $\varrho_1 < \varrho_2$. Conversely, the decision in favor of hypothesis 2 is made when $\varrho_1 > \varrho_2$, which is indicated using the following notation:

$$\varrho_1 \underset{\delta_2}{\overset{\delta_1}{\lessgtr}} \varrho_2 \qquad (2.19)$$

The above can be rewritten using Equations (2.17) and (2.18) to

$$\frac{p_2}{p_1} \underset{\delta_1}{\overset{\delta_2}{\gtrless}} \frac{L_{12} - L_{11}}{L_{21} - L_{22}} \tag{2.20}$$

In the above Equation (2.20) it was assumed that the loss due to incorrect decisions is higher than the loss incurred by correct decisions, i.e., $L_{12} > L_{11}$ and $L_{21} > L_{22}$ which must be true for rational decisions. From the above it is evident that the Bayesian binary hypothesis testing is based on a simple comparison of the ratio of posterior probabilities to some threshold which is dependent on the loss function.

To develop more intuition about the Bayesian hypothesis testing we use the fact that $p_1 = 1 - p_2$. Assuming that the loss function for correct decisions is zero, which is almost always a reasonable assumption in any decision problem, and denoting the ratio $L_{12}/L_{21} = x$, Equation (2.20) can be further transformed using Equations (2.17) and (2.18) to

$$p_2 \underset{\delta_1}{\overset{\delta_2}{\gtrless}} \frac{x}{1+x}. \tag{2.21}$$

For the Bayesian binary hypothesis testing it is therefore sufficient to compute the posterior of one of the hypotheses and make a decision based on comparison of this posterior probability to a threshold derived from the loss function.

In the remainder of this section we will show equivalence of posterior-expected-loss decision principle with the so-called *ideal observer* [6] ofter used in the literature. Before defining the ideal-observer decision principle, a new unobservable quantity (UQ) of interest is defined. We slightly deviate from the convention of this chapter in which we indicate all uncertain quantities with θ. It is done to achieve compatibility in notation with other texts containing derivation of the ideal observer.

> **Notation:** We define the *hypothesis* as new quantity of interest and denote it as H. In the binary hypothesis testing it can have only 2 values and in our case it is 1 or 2. We will use notation such $H_1 \equiv H = 1$ and $H_2 \equiv H = 2$.

We assume that if we know $\boldsymbol{\theta}$ (true value of UQs), we know H (which hypothesis is true or not and therefore) which implies that if $\boldsymbol{\theta}$ is uncertain then the H is uncertain. To put it formally:

$$p(H_1|\boldsymbol{\theta}) = \begin{cases} 1 \text{ if } \boldsymbol{\theta} \in \boldsymbol{\Theta}_1 \\ 0 \text{ if } \boldsymbol{\theta} \notin \boldsymbol{\Theta}_1 \end{cases} \tag{2.22}$$

and similarly for $p(H_2|\boldsymbol{\theta})$. As before $\boldsymbol{\Theta} = \boldsymbol{\Theta}_1 \cup \boldsymbol{\Theta}_2$ and $\boldsymbol{\Theta}_1 \cap \boldsymbol{\Theta}_2 = \emptyset$.

Example 2.10: Conditional two-dice roll (9)

In Example 2.9 hypothesis H_1 was specified as the proposition that the number on the first roll θ is either 1, 2, 3, or 4, and hypothesis H_2 that the number is 5 or 6. Therefore, $p(H_1|\theta = 1) = 1$, $p(H_1|\theta = 5) = 0$, $p(H_2|\theta = 6) = 1$, etc.

Using Equation (2.22) and Bayes Theorem we obtain

$$p(\boldsymbol{\theta}) = \begin{cases} p(\boldsymbol{\theta}|H_1)p(H_1) & \text{if } \boldsymbol{\theta} \in \boldsymbol{\Theta}_1 \\ p(\boldsymbol{\theta}|H_2)p(H_2) & \text{if } \boldsymbol{\theta} \in \boldsymbol{\Theta}_2 \end{cases} \tag{2.23}$$

The posterior expected loss rule for decision making was given by Equation (2.20) and is

$$\frac{p_2}{p_1} \underset{\delta_1}{\overset{\delta_2}{\gtrless}} \frac{L_{12} - L_{11}}{L_{21} - L_{22}} \tag{2.24}$$

since $p_{1,2}$ were defined as $p_{1,2} = \int_{\boldsymbol{\theta} \in \boldsymbol{\Theta}_{1,2}} p(\boldsymbol{\theta}|\mathbf{G} = \mathbf{g})$ we have

$$\frac{p_2}{p_1} = \frac{\int_{\boldsymbol{\theta} \in \boldsymbol{\Theta}_2} p(\boldsymbol{\theta}|\mathbf{G} = \mathbf{g})}{\int_{\boldsymbol{\theta} \in \boldsymbol{\Theta}_1} p(\boldsymbol{\theta}|\mathbf{G} = \mathbf{g})} = \frac{\int_{\boldsymbol{\theta} \in \boldsymbol{\Theta}_2} p(\mathbf{G} = \mathbf{g}|\boldsymbol{\theta})p(\boldsymbol{\theta})}{\int_{\boldsymbol{\theta} \in \boldsymbol{\Theta}_1} p(\mathbf{G} = \mathbf{g}|\boldsymbol{\theta})p(\boldsymbol{\theta})}. \tag{2.25}$$

The right-hand side of the above was obtained using the Bayes Theorem. Using now Equation (2.23) we obtain

$$\frac{p_2}{p_1} = \frac{p(H_2) \int_{\boldsymbol{\theta} \in \boldsymbol{\Theta}_2} p(\mathbf{G} = \mathbf{g}|\boldsymbol{\theta})p(\boldsymbol{\theta}|H_2)}{p(H_1) \int_{\boldsymbol{\theta} \in \boldsymbol{\Theta}_1} p(\mathbf{G} = \mathbf{g}|\boldsymbol{\theta})p(\boldsymbol{\theta}|H_1)} = \frac{p(H_2) \int_{\boldsymbol{\theta} \in \boldsymbol{\Theta}} p(\mathbf{G} = \mathbf{g}|\boldsymbol{\theta})p(\boldsymbol{\theta}|H_2)}{p(H_1) \int_{\boldsymbol{\theta} \in \boldsymbol{\Theta}} p(\mathbf{G} = \mathbf{g}|\boldsymbol{\theta})p(\boldsymbol{\theta}|H_1)} \tag{2.26}$$

Note that the integration limits were changed to the entire set $\boldsymbol{\Theta}$. This is because in the numerator for $\boldsymbol{\theta} \in \boldsymbol{\Theta}_1$: $p(\boldsymbol{\theta}|H_2) \propto p(H_2|\boldsymbol{\theta}) = 0$ and similarly in the denominator. Because hypotheses H are conditionally independent on $\mathbf{g}$ given $\boldsymbol{\theta}$, we have that $p(\mathbf{G} = \mathbf{g}|\boldsymbol{\theta}) = p(\mathbf{G} = \mathbf{g}|\boldsymbol{\theta}, H)$ and therefore

$$\frac{p_2}{p_1} = \frac{p(H_2) \int_{\boldsymbol{\theta} \in \boldsymbol{\Theta}} p(\mathbf{G} = \mathbf{g}|\boldsymbol{\theta}, H_2)p(\boldsymbol{\theta}|H_2)}{p(H_1) \int_{\boldsymbol{\theta} \in \boldsymbol{\Theta}} p(\mathbf{G} = \mathbf{g}|\boldsymbol{\theta}, H_1)p(\boldsymbol{\theta}|H_1)} = \frac{p(H_2)}{p(H_1)} \frac{p(\mathbf{G} = \mathbf{g}|H_2)}{p(\mathbf{G} = \mathbf{g}|H_1)}. \tag{2.27}$$

The ratio *likelihood ratio* Λ is defined as

$$\Lambda(\mathbf{G} = \mathbf{g}) = \frac{p(\mathbf{G} = \mathbf{g}|H_2)}{p(\mathbf{G} = \mathbf{g}|H_1)} \tag{2.28}$$

and using Equations (2.24), (2.27) and (2.28) we transform the posterior expected loss decision rule to

$$\Lambda(\mathbf{G} = \mathbf{g}) \underset{\delta_1}{\overset{\delta_2}{\gtrless}} \frac{p(H_1)(L_{12} - L_{11})}{p(H_2)(L_{21} - L_{22})} \tag{2.29}$$

The above is the ideal-observer (IO) rule of decision making used in imaging.

The Bayesian posterior expected loss rule given by Equation (2.16) is the general decision rule in any hypothesis testing. The IO is derived from this rule for the binary hypothesis testing and is defined in terms of generalized Bayesian likelihoods and is a specific formulation of a more general Bayesian decision making defined by the lowest expected loss function value. We strongly prefer the simpler formulation given in Equations (2.2) and (2.16).

The IO decision rule is based on the *likelihood ratio* which is a quite unfortunate name because it is misleading as to the nature of the decision rule. The actual rule, as demonstrated, is derived from the fully Bayesian decision rule; nevertheless, the name may suggest that it is based on the ratio of likelihoods which is typically not true. Only in extreme cases when simple hypotheses are considered for which there is one-to-one correspondence between hypotheses and θ, then the likelihood ratio Λ becomes the actual ratio of likelihoods. This is, however, a radical simplification where there are only two possible values of θ and has a little value in practical applications.

Example 2.11: Conditional two-dice roll (10)

In continuation of the example with two dice, we calculate various new probabilities and quantities as the Bayesian likelihood ratio (LR). Since the LR is a function of data (observed quantity **g**) it is then an observable quantity as well, and in AE condition the value is known. In our example the unobservable quantity (UQ) is the number obtained in the first roll. The prior probability of obtaining each number $1, \ldots, 6$ in roll 1 is $p(1) =, \ldots, = p(6) = 1/6$. We use the notation that $p(\theta = 1) = p(1)$ etc. H_1 corresponds to the hypothesis that number 1, 2, 3, or 4 was obtained in the first roll and H_2 corresponds to the hypothesis that the number 5 or 6 was obtained. It follows that the probability of hypothesis H_1 is $p(H_1) = p(1) + p(2) + p(3) + p(4) = 4/6$ and similarly $p(H_2) = 2/6$. The conditional $p(1|H_1) = p(1)/p(H_1) = 1/4$ and similarly $p(2|H_1) = p(3|H_1) = p(4|H_1) = 1/4$. As before in Example 2.9 the loss function is defined as $L_{11} = L_{22} = 0$ and $L_{12} = L_{21} = x$. Using these numbers the decision threshold for IO is calculated from Equation (2.29) as

$$\frac{p(H_1)(L_{12} - L_{11})}{p(H_2)(L_{21} - L_{22})} = \frac{\frac{4}{6}x}{\frac{2}{6}x} = 2 \qquad (2.30)$$

If the value of the likelihood ratio is above 2, the H_2 is selected and if it is below 2 the H_1 is selected. The computation of the likelihood ratio is shown in Table 2.6. In the calculation, the values of the likelihood $p(G = g|\theta)$ were used from Table 1.1. The optimal decisions are the same as obtained using posterior expected loss (Example 2.9) which should not be surprising as that likelihood ratio test was derived from the general Bayesian principle used in Example 2.9. We needed to go through quite a few hurdles in order to compute all required probabilities in order to obtain the likelihood ratio even though a simple example was used here. In general, the computation of the Bayesian likelihood ratio is quite challenging. It is important to keep in mind that since Λ represents a ratio, sometimes a division by zero will occur. Even the decision threshold can be $\pm\infty$. These possibilities create potential problems in numerical stability of computation of the likelihood ratio test and make this approach less attractive for use in Bayesian computing than methods based on the posterior expected loss.

TABLE 2.6

Generalized Bayesian likelihoods and likelihood ratios

		$p(G = g\|H_1)$	$p(G = g\|H_2)$	$\Lambda(G = g)$ $= \frac{p(G=g\|H_2)}{p(G=g\|H_1)}$	Best decision (decision threshold = 2)
$g \downarrow$	1	$25/48^a$	$11/60$	$44/125^b$	H_1
	2	$13/48$	$11/60$	$44/65^b$	H_1
	3	$7/48$	$11/60$	$44/35^b$	H_1
	4	$3/48$	$11/60$	$44/15^c$	H_2
	5	0	$11/60$	∞^c	H_2
	6	0	$1/12$	∞^c	H_2

^{a}Example of calculation: From Equation (2.27) we have that
$p(g = 1|H_1) = p(g = 1|1)p(1|H_1) + \ldots + p(g = 1|6)p(6|H_1)$. The likelihoods from Table 1.1 are $1, \frac{1}{2}, \frac{1}{3}, \frac{1}{4}, \frac{1}{5}, \frac{1}{6}$ for $p(g = 1|1)$ through $p(g = 1|6)$, respectively. Therefore, $p(g = 1|H_1) = 1 \times \frac{1}{4} + \frac{1}{2} \times \frac{1}{4} + \frac{1}{3} \times \frac{1}{4} + \frac{1}{4} \times \frac{1}{4} + \frac{1}{5} \times 0 + \frac{1}{6} \times 0 = 25/48$.
^{b}The value is less than the decision threshold 2 and therefore decision in favor of H_1 is made.
^{c}The value is larger than the decision threshold 2 and therefore H_2 is decided.

2.4 BEFORE-THE-EXPERIMENT DECISION MAKING

First, we explain why we would ever want to make a decision before seeing the data (realizations of OQs). That seems unreasonable because upon seeing the data more information about the UQs may be available and OQs are known. The reason for this is that BE decision making will be used for evaluation of decision principles. For example, we will be interested in whether δ_1 is better than δ_2. Obviously, we can compare those decision principles after the experiment by computing the posterior expected loss but such comparison would rely on a particular data set and there would be no guarantee that the same conclusion is reached when different data is observed.

An example of application of BE decision making in medical imaging is a situation in which it has to be decided whether one imaging scanner is better than another one for some medically indicated task. BE decision making will also be useful for evaluation of image processing methods.

How well the task is performed will be assessed by the *expected loss* function. There is a slight difference in terminology here because before we used the posterior expected loss and now we use the expected loss as the expected loss is not tied to a particular posterior and is calculated before the experiment so the posterior is not known[4]. To avoid confusion, we will use a commonly used name for the expected risk in BE state, the *Bayes risk*. In the BE decision making we will use the joint probability of all QoIs to compute the expected loss, because all QoIs are uncertain.

Therefore, the $p^*(\varphi)$ in Equation (2.1) is replaced with the joint of $p(\boldsymbol{\theta}, \mathbf{g})$ where by the convention adopted in this chapter $\boldsymbol{\theta}$ corresponds to all unobservable quantities (UQs) and $\mathbf{g}$ corresponds to observable quantities. Denoting Bayes risk by ϖ and using Equation (2.1) we obtain

$$\varpi(\delta) = \int_{\mathbf{g}\in\mathbf{G}} \int_{\boldsymbol{\theta}\in\boldsymbol{\Theta}} L(\boldsymbol{\theta}, \delta^{\mathbf{g}}) p(\boldsymbol{\theta}, \mathbf{g}) \tag{2.31}$$

the above can be rewritten as

$$\varpi(\delta) = \int_{\mathbf{g}\in\mathbf{G}} p(\mathbf{g}) \int_{\boldsymbol{\theta}\in\boldsymbol{\Theta}} L(\boldsymbol{\theta}, \delta^{\mathbf{g}}) p(\boldsymbol{\theta}|\mathbf{g}) \tag{2.32}$$

Equation (2.32) has a form that clearly shows the nature of the decision process in the BE stage. The term $\int_{\boldsymbol{\theta}\in\boldsymbol{\Theta}} L(\boldsymbol{\theta}, \delta^{\mathbf{g}}) p(\boldsymbol{\theta}|\mathbf{g})$ is the posterior expected loss assuming the $\mathbf{g}$ is observed (see the definition in Equation (2.2)). Therefore, all hypothetical values of posterior expected losses are averaged and weighted by the probability $p(\mathbf{g})$ that the data is obtained. The value of this average is the figure of merit of the decision principle δ (Bayes risk). The $\varpi(\delta)$ is the Bayes risk (expected loss) of the decision principle δ which is independent on $\mathbf{g}$.

The Bayes risk (Equation (2.32)) can also be rewritten in an alternative form

$$\varpi(\delta) = \int_{\boldsymbol{\theta}\in\boldsymbol{\Theta}} p(\boldsymbol{\theta}) \int_{\mathbf{g}\in\mathbf{G}} L(\boldsymbol{\theta}, \delta^{\mathbf{g}}) p(\mathbf{g}|\boldsymbol{\theta}) \tag{2.33}$$

This formulation of the decision risk is dependent on the statistical model $p(\mathbf{g}|\boldsymbol{\theta})$. In fact, $\int_{\mathbf{g}\in\mathbf{G}} L(\boldsymbol{\theta}, \delta^{\mathbf{g}}) p(\mathbf{g}|\boldsymbol{\theta})$ is the classical *risk function* used in frequentist statistics for decision making [8]. Since the classical risk is used this decision principle is not fully Bayesian (albeit the name) as the fully Bayesian analyses are always done in AE conditions using the posterior [8]. We will not get into a discussion about the classical vs. Bayesian decision principles and the interested reader is encouraged to consult the classic texts such Berger [8], Gelman et al. [31], Lehmann and Romano [67], and Robert [84].

---

[4] As it was explained in Chapter 1, both the posterior and the likelihood are defined in AE stage when the data are known.

2.4.1   BAYES RISK

By examining Equation (2.32) it should be obvious that if $\delta^{\mathbf{g}}$ is the Bayesian decision rule, the Bayes risk is minimized for that rule. This is because from the definition $\int_{\boldsymbol{\theta} \in \Theta} L(\boldsymbol{\theta}, \delta^{\mathbf{g}}) p(\boldsymbol{\theta}|\mathbf{g})$ is at its minimum for Bayesian decision rule and therefore the result of integration (or summation for discrete $\mathbf{g}$) is minimized as well. It follows that point estimators that were developed for AE decision making in previous sections will also minimize Bayes risk and therefore they will be the best point estimation procedures available as well. In fact all Bayesian decision making rules based on the posterior expected loss are also optimal in terms of Bayes risk and therefore are optimal from the decision-theoretic point of view. To illustrate these concepts let's consider the following example:

Example 2.12: Conditional two-dice roll (11)

We continue Example 2.4.

Person A rolls the first die obtaining result θ and then rolls the second die obtaining result g until the number is equal or less than the result of the first roll. Only the final number obtained on the second die g is revealed to the observer.

We determined the MMSE estimators (one for each data) which are given in Table 2.4. In the current example we compute the expected loss (Bayes risk) of the MMSE estimator and compare it to the expected loss of MAP estimator and to an ad hoc estimator specified below. Before we specify the ad hoc estimator we point out that MAP estimator will not minimize the Bayes risk because it will be evaluated under the quadratic loss function and therefore we expect the Bayes risk of this estimator to be higher than for the MMSE. The exercise here is to confirm if this is in fact true.

Ad hoc estimator: Based on the description of the experiment, it seems reasonable to assume that the value on the first roll may be somewhere in the middle between the number obtained and the maximum number that could have been obtained on the first roll (which is 6) and therefore we define it as

$$\hat{\theta}_{AH}(g) = \frac{g+6}{2} \tag{2.34}$$

For the problem considered here we can either use Equations (2.32) and (2.33) or calculate the Bayes risk. More often for practical application, the form of Equation (2.33) will be used as it requires the prior $p(\boldsymbol{\theta})$ and model $p(\mathbf{g}|\boldsymbol{\theta})$ which are more readily available and easier to obtain than for example $p(\mathbf{g})$ needed for calculation of Bayes risk using Equation (2.32). We show the calculation using both approaches (expecting they will give identical results) and present the findings in Table 2.7.

The Bayes risk is optimized for the Bayesian point estimators under the assumption that the estimator uses the perfect prior $p(\boldsymbol{\theta})$. How do we define

TABLE 2.7

Bayes risk under quadratic loss function computed using Equations (2.32) and (2.33)

$$\varpi(\hat{\theta}) = \int_{g \in G} p(g) \int_{\theta \in \Theta} L(\theta, \hat{\theta}))p(\theta|g) \quad \| \quad \varpi(\hat{\theta}) = \int_{\theta \in \Theta} p(\theta) \int_{g \in G} L(\theta, \hat{\theta})p(g|\theta)$$

g	$p(g)$	Value of $\int_{\theta \in \Theta} L(\theta, \hat{\theta})p(\theta\|g)$			θ	$p(\theta)$	Value of $\int_{g \in G} L(\theta, \hat{\theta})p(g\|\theta)$		
		$\hat{\theta}_{\mathrm{MMSE}}$	$\hat{\theta}_{\mathrm{MAP}}$	$\hat{\theta}_{\mathrm{AH}}$			$\hat{\theta}_{\mathrm{MMSE}}$	$\hat{\theta}_{\mathrm{MAP}}$	$\hat{\theta}_{\mathrm{AH}}$
1	147/360	2.57[a]	4.67	3.68	1	1/6	2.10[b]	0.00	6.25
2	87/360	1.90	4.00	2.21	2	1/6	1.15	0.50	3.12
3	57/360	1.21	2.68	1.30	3	1/6	0.66	1.67	1.17
4	37/360	0.66	1.41	0.68	4	1/6	0.88	3.50	0.37
5	22/360	0.25	0.45	0.25	5	1/6	1.95	6.00	0.75
6	10/360	0.00	0.00	0.00	6	1/6	3.98	9.17	2.29
	$\varpi(\hat{\theta})$	1.79[c]	3.47	2.33		$\varpi(\hat{\theta})$	1.79	3.47	2.33

[a] This value can be interpreted as the average loss when $g = 1$ is observed and MMSE estimator is used.

[b] This value is interpreted as average loss when the true number on die A is 1 when using MMSE estimator. Similar interpretations applied to other values in the table.

c This is expected loss (Bayes risk) when using the MMSE estimator.

Note: As expected, the Bayes risk of MMSE estimator is the lowest. Interestingly, the ad hoc (AH) estimator is better than MAP estimator when quadratic loss function was used.

the perfect prior? One was to say that if a perfect prior is used and perfect model is used then the true values are samples from the resulting posteriors when an identical experiment is repeated [19]. This interpretation is very much frequentist in nature. Suffice to say, the knowledge of perfect priors is difficult to obtain and often an approximate prior has to be used. Here we would like to emphasize that considerations of inaccurate priors make sense only in BE. For the fully Bayesian treatment based on the posteriors in AE, this problem is undefined as the prior indicates the current knowledge (BE) about the UQ and consideration of correctness of these priors are irrelevant for the correctness of Bayesian analysis.

In BE regime, with the Bayes risk we can quantify the effect of inaccurate priors. Suppose we want to evaluate an imaging apparatus and data processing methods (which we call *imaging system*) using computer simulations. The outcome of the imaging system is a point estimate of the value of the blood flow in the heart denoted by θ (see also Chapter 6 for other examples). For now, the details of how this measurement can be done are irrelevant. To perform such simulation we assume what population of patients will be scanned by this imaging system and describe this population by the distribution $p(\theta)$. Such

$p(\theta)$ describes what we call real uncertainty about the θ which in practical applications will never be known. Ideally the Bayes risk is calculated by using such prior when making the decision.

Suppose now that data are obtained and since the $p(\theta)$ is not known exactly the posterior is formed assuming some prior $\pi(\theta)$ which does not necessarily have to be the same as $p(\theta)$. The posterior derived from that prior is then used to determine the point estimator $\hat{\theta}^{\pi}$. The Bayes risk with such mismatched priors will be

$$\varpi(\hat{\boldsymbol{\theta}}^{\pi}) = \int_{\boldsymbol{\theta} \in \boldsymbol{\Theta}} p(\boldsymbol{\theta}) \int_{\mathbf{g} \in \mathbf{G}} L(\boldsymbol{\theta}, \hat{\boldsymbol{\theta}}^{\pi}) p(\mathbf{g}|\boldsymbol{\theta}). \tag{2.35}$$

It reflects not only the performance of the estimator, but also the effect of the mismatch between $p(\theta)$ and $\pi(\theta)$.

An extreme case of such approach is when the performance of the imaging system is investigated for $p(\boldsymbol{\theta}) = \delta_d(\boldsymbol{\theta})$[5] where δ_d is the Kronecker or Dirac delta (depending on $\boldsymbol{\theta}$ being discrete or continuous). This is equivalent to requiring that the UQ that will be subject to imaging can only have single value specified by such prior. Although this seems like a gross simplification, it is a common assumption made by many investigators in the imaging field when methods are tested using a single object. When assuming that $p(\boldsymbol{\theta}) = \delta_d(\boldsymbol{\theta})$ the Bayes risk becomes equivalent to the mean square error (MSE)[6] measure under the quadratic loss function. This can be shown as follows:

$$\varpi(\hat{\boldsymbol{\theta}}^{\pi}) = \int_{\boldsymbol{\theta}' \in \boldsymbol{\Theta}} \delta_d(\boldsymbol{\theta}) \int_{\mathbf{g} \in \mathbf{G}} L(\boldsymbol{\theta}', \hat{\boldsymbol{\theta}}^{\pi}) p(\mathbf{g}|\boldsymbol{\theta}') = \int_{\mathbf{g} \in \mathbf{G}} (\boldsymbol{\theta} - \hat{\boldsymbol{\theta}}^{\pi})^T (\boldsymbol{\theta} - \hat{\boldsymbol{\theta}}^{\pi}) p(\mathbf{g}|\boldsymbol{\theta})$$
$$\tag{2.36}$$

The right-hand side of this equation is the definition of classical mean square error [57]. This widely popular approach to evaluation of imaging system and algorithms is a special case of Bayes risk obtained after a quite drastic simplification of the definition of Bayes risk ϖ. In Example 2.13 we show some interesting phenomena associated with the use of MSE and the use of mismatched prior which is the violation of *admissibility* of Bayes rule. The issue of admissibility of the estimator (or any other statistical rule) arises when the estimator is not guaranteed to be optimal at the entire range of values of $\boldsymbol{\theta}$. This is demonstrated in Example 2.13.

---

[5]There is a notation conflict because in the same chapter we use δ to denote a Bayesian decision as well as here to indicate the Kronecker or Dirac delta. To distinguish these two, we use subscript δ_d to indicate the Kronecker/Dirac delta.

[6]Although there is a high similarity in name of the minimum-mean-square-error (MMSE) estimator which is equal to the mean of the posterior and defined here mean square error (MSE) which is the name of the error measure, the MMSE and MSE are not related and there is no meaningful connection between them.

Example 2.13: Conditional two-dice roll (12)

We reexamine the three estimators that we investigated before. However, here we introduce a mismatch between the prior for computing the Bayes risk in Equation (2.33) and the prior used for the calculation of the estimator. As before, suppose that at the time of decision making we believe that dice are fair and therefore we use flat prior $\pi(\theta) = 1/6$ for any value of $\theta = 1, \ldots, 6$. However, the die A is loaded such that the chance of obtaining 6 is in fact much higher and equal to 2/6, and the chance of obtaining 1 is reduced to 0. All probabilities of numbers other than 1 and 6 remain the same and are equal to 1/6. The results of calculation of Bayes risk for this case are shown in the left half of Table 2.8.

We also modified the prior $p(\theta)$ into a Kronecker delta. Therefore, in the experiment the result of roll A is always the same number but the decision maker uses the flat prior (equal probability assigned to each number). The Bayes risk is equal to MSE error in such case. Depending on the value of θ different values of MSE are obtained and shown in rows of the right half of Table 2.8. It is evident that depending on θ different estimators are optimal (have minimum MSE) and therefore none of them is admissible.

TABLE 2.8
Bayes risk for loaded dice and MSE computed using Equation (2.33)

$$\varpi(\hat{\theta}) = \int_{\theta \in \Theta} p(\theta) \int_{g \in G} L(\theta, \hat{\theta}) p(g|\theta) \quad \| \quad \mathrm{MSE} = \int_{\mathbf{g} \in \mathbf{G}} (\boldsymbol{\theta} - \hat{\boldsymbol{\theta}}^{\pi})^T (\boldsymbol{\theta} - \hat{\boldsymbol{\theta}}^{\pi}) p(\mathbf{g}|\boldsymbol{\theta})$$

		Value of $\int_{g \in G} L(\theta,\hat{\theta})p(g\|\theta)$				Value of MSE			
θ	$p(\theta)$	$\hat{\theta}_{\mathrm{MMSE}}$	$\hat{\theta}_{\mathrm{MAP}}$	$\hat{\theta}_{\mathrm{AH}}$	θ	$\hat{\theta}_{\mathrm{MMSE}}$	$\hat{\theta}_{\mathrm{MAP}}$	$\hat{\theta}_{\mathrm{AH}}$	Best[a]
1	0	2.57	4.67	3.68	1	2.10	0.00	6.25	MAP
2	1/6	1.90	4.00	2.21	2	1.15	0.50	3.12	MAP
3	1/6	1.21	2.68	1.30	3	0.66	1.67	1.17	MMSE
4	1/6	0.66	1.41	0.68	4	0.88	3.50	0.37	AH
5	1/6	0.25	0.45	0.25	5	1.95	6.00	0.75	AH
6	2/6	0.00	0.00	0.00	6	3.98	9.17	2.29	AH
$\varpi(\hat{\theta})$		2.10	5.00	1.67					

[a] This is the estimator with the lowest MSE for given value of θ. The estimator with the lowest value in each row is selected.

Note: For the case of leaded die A the ad hoc estimator is the best (shown in left part of the table). Depending on the value of θ, different estimators are optimal and therefore neither of them is admissible.

2.4.2 OTHER METHODS

The Bayes risk discussed in the previous section uses the loss function. In this section we briefly introduce two other approaches that can be used in BE condition for system optimization (experimental design) that are based on computation of some *utility function* that quantifies the difference between the prior and the posterior. Since the loss function is not utilized it is not optimal from the decision-theoretic perspective. However, these two methods are quite straightforward to implement and are useful for situations where the loss function is unknown or cannot be reliably specified.

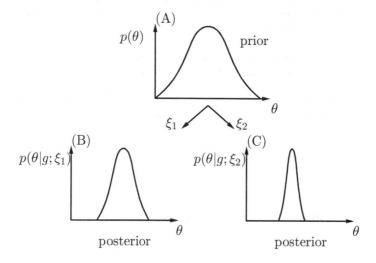

FIGURE 2.5 Example in 1D of two different imaging systems (ξ_1 and ξ_2) t generate two different posteriors. Width of those posteriors is indicative of how well the system performs.

The main idea of these approaches is summarized in Figure 2.5. The "wider" the posterior, there is less knowledge about unobservable quantities. To quantify this we define *utility function* or simply *utility* to indicate the effect of the data on the "width" of the posterior. We use quotation marks because the width is well defined in 1D, but the analysis can be applied in any dimensions in which case the width may now accurately describe the effect. The utility function can be considered as the opposite of the loss function. Therefore, the higher the value of utility the better the experimental design (imaging system).

Before we define the utility function, the *Shannon information entropy*

[20, 81, 91] is defined as

$$\Xi = - \int_{\theta \in \Theta} p(\theta) \log p(\theta) \tag{2.37}$$

where $p(\theta)$ is a probability distribution of quantity θ and $0 \log 0$ is assumed to be 0 because $\lim_{x \to 0} x \log x = 0$. The $p(\theta)$ could either be continuous or discrete. The Shannon entropy was developed for the theory of communication as an indication of the amount of information contained in a message. The higher Shannon entropy indicates the larger amount of information. For our application the Shannon entropy is used for a different task and we need to be careful not to confuse the amount of information used by Shannon in the communication theory with the amount of knowledge about UQs. For our use, the Shannon entropy is a measure of unpredictability expressed by the probability distribution. Systems that achieve lower entropy (less unpredictability) will be preferable over systems that achieve the higher entropy (more unpredictability). In Figure 2.5 the posterior shown in (B) has a higher Shannon entropy than the posterior shown in (D) and therefore will be less preferable.

We define the utility U as a function of observed data $\mathbf{g}$ and imaging system ξ used to acquire and process the data as the difference in *Shannon entropy* between the prior and the posterior. It is expected that the prior distribution has a higher entropy because after the data is acquired the uncertainty about UQs should be reduced. In other words we expect that the entropy of the prior will be higher than the entropy of the posterior. Mathematically the utility function is

$$U(\mathbf{g}; \xi) = - \int_{\theta \in \Theta} p(\theta) \log p(\theta) - \left(- \int_{\theta \in \Theta} p(\theta|\mathbf{g}; \xi) \log p(\theta|\mathbf{g}; \xi) \right) \tag{2.38}$$

Using the above definition, the average difference in Shannon entropy between prior and posterior [69] is

$$\bar{U}(\xi) = \int_{\mathbf{g} \in G} p(\mathbf{g}; \xi) U(\mathbf{g}; \xi)$$

$$= \int_{\mathbf{g} \in G} \int_{\theta \in \Theta} p(\theta, \mathbf{g}; \xi) \log p(\theta|\mathbf{g}; \xi) - \int_{\theta \in \Theta} p(\theta) \log p(\theta) \tag{2.39}$$

Another formulation of the utility is obtained using the Kullback–Leibler divergence (KL divergence) between posterior and prior. The KL divergence quantifies the difference between two probability distributions and therefore a higher divergence will indicate a better imaging system. The KL divergence between posterior and prior is defined as

$$D_{\mathrm{KL}} \left(p(\theta|\mathbf{g}) \,||\, p(\theta) \right) = \int_{\theta \in \Theta} p(\theta|\mathbf{g}) \log \frac{p(\theta|\mathbf{g})}{p(\theta)}. \tag{2.40}$$

The KL divergence is non-negative and it is zero only when $p(\theta|\mathbf{g})$ and $p(\theta)$ are identical. As in the case of the Shannon entropy $0 \log 0$ is considered as equal to 0. If the $p(\boldsymbol{\theta})$ in denominator in Equation (2.40) is zero, it implies (from Bayes theorem) that $p(\boldsymbol{\theta}|\mathbf{g})$ is zero and the expression Equation (2.40) evaluates to zero. The average utility using the KL divergence is

$$\bar{U}_{KL}(\xi) = \int_{\mathbf{g} \in \mathbf{G}} p(\mathbf{g}; \xi) D_{\mathrm{KL}}\left(p(\boldsymbol{\theta}|\mathbf{g}; \xi) \,||\, p(\boldsymbol{\theta})\right) =$$

$$\int_{\mathbf{g} \in \mathbf{G}} \int_{\boldsymbol{\theta} \in \boldsymbol{\Theta}} p(\boldsymbol{\theta}, \mathbf{g}; \xi) \log \frac{p(\boldsymbol{\theta}|\mathbf{g}; \xi)}{p(\boldsymbol{\theta})} \quad (2.41)$$

Other measures that indicate the uncertainty in the posterior like the posterior variance [96] can also be used.

2.5   ROBUSTNESS OF THE ANALYSIS

The last section of this chapter is important for anyone who wants to effectively practice Bayesian statistics. In analysis of phenomena that involve uncertainty, there is no objective methods that would provide certain answers to questions asked. There are always assumptions that need to be made and some level of subjectivity that needs to be used.

Even if we make assumptions about the model of the experiment, it is important that small deviations in those assumptions do not dramatically change the conclusion of the investigation. The analysis of this sensitivity to the assumptions in Bayesian analysis is called the *robustness analysis* or *sensitivity analysis*. In Bayesian analysis three sources of subjectivity and associated robustness of the approach to those need to be considered:

Selection of statistical model. The statistical models describe the rules of nature and in general they may not be exactly known. The nature is much more complex than it can be modeled in any theoretical or numerical experiment and therefore certain simplifications of the model are always used. The selection of the model can be considered as a particular type of prior specification. This issue is important not only for Bayesian but for anyone who investigates uncertain phenomena. In this book we will not be concerned with uncertain statistical models. For specific examples and discussion on model selection for Bayesian analysis see Berger [8], Box [11], and Box and Tiao [13].

Selection of the loss function. Interestingly, in the Bayesian decision theory, the specification of the loss function may be even more uncertain than specification of the prior. However, the specification of the loss function does not suffer from the robustness problems as does the specification of the prior. The reason for this is that in specification of the prior, the beliefs about very low probable values of quantities are difficult to specify (tails of distributions) and are the frequent source of lack of robustness [8], whereas the loss function is "applied" after seeing the data and therefore the tails of the posterior will not be as important (the posterior is much narrower than the prior). It follows that the tails of the loss are much less important which makes it easier to specify the robust loss function.

Selection of the prior. The most important and a source of much debate is the robustness of the prior. If changing our prior beliefs changes considerably the outcome of the analysis, we should rethink the approach to the problem and redesign the experiment. In general it is difficult to perform formal robustness analysis of the prior [8]. However, any Bayesian investigation should contain some analysis of the robustness. Even as simple as trial-and-error approach is desirable in which the prior is slightly modified and the effect on the result of the analysis is examined. Almost from the definition, uninformative priors are robust because they do not bring much information to the analysis and therefore a slight change will not change the result.

3 Counting statistics

3.1 INTRODUCTION TO STATISTICAL MODELS

In this chapter the basic laws that govern stochastic processes in nuclear imaging when using photon-limited data (PLD) are introduced. The PLD consist of discrete events (e.g., radioactive decays) that occur at some locations within the object. These events may or may not be detected/registered by the medical imaging detector system (the camera). Detectors capable of registering discrete events will be referred to as "photon-counting detectors" (see Chapter 5 where some of such detectors are discussed).

Counting statistics defined in this chapter describe the "laws of nature" that govern occurrence and detection of discrete events. These laws will serve as the basis for statistical model definitions. As explained in Chapter 1, models are necessary for obtaining likelihoods, posteriors, and pre-posteriors which are the main tools for all decision tasks discussed. Although authors of many texts [6, 17, 18, 38, 39] define counting statistics following Poisson or Gaussian (also referred to as Normal) distribution they are approximations of more general laws. The general models used here are based on binomial, multinomial, and sums of not identically distributed binomial and multinomial distributions. The approximations that lead to the Poisson and normal distributions are demonstrated.

The true nature of PLD in nuclear imaging is based on the fact that the unobservable quantities (UQs) are discrete and observable quantities (OQs) are discrete. For example, in emission tomography there are radioactive nuclei that are distributed in the object. The number of the nuclei is discrete and unknown and considered as UQ. This number constitute the state of nature (SoN). When photon counting detectors are used, the discrete number of counts obtained by those detectors will be the QoI that is observed (OQ). Therefore, the definition of the model is equivalent to finding laws that describe the number of detections given the number of radioactive nuclei. It is clear that all QoI are discrete and therefore a discrete-to-discrete (D-D) model describes the statistics of such processes. When nuclear imaging is concerned the D-D model is always the general model of the state of nature (SoN).

The approximation that is used when D-D data are analyzed is to assume that UQs are continuous. It is a reasonable assumption when the numbers of discrete events in the imaged objects are very high. It also makes the computing somewhat easier. For example, when working with continuous quantities, the differentiation operation becomes available which is a part of many optimization and data analysis algorithms. In nuclear imaging when continuous model of UQs is used and a discrete-continuous (D-C) statistical model is defined. In this model D stands for discrete data and C stands for continues model. Application of this approximation leads to the Poisson distribution.

The final type of the statistical model that will be discussed is a continuous-to-continuous (C-C) model described by the normal distribution. This model is derived in this book directly from the D-D model but it is also possible to obtain the model from D-C model governed by Poisson statistics.

Our hope is that the presentation of the statistics with various levels of approximations will allow us to fully appreciate the basics of the statistical description of the PLD imaging processes and also understand the nature of the Poisson or normal distribution approximations. We provide models and approximations leading to D-C and C-C models for completeness. However, the computational algorithms discussed later in this book (Chapter 6) use the general D-D models.

Figure 3.1 presents the layout of this chapter and shows interconnections between the different formulations of the statistical laws.

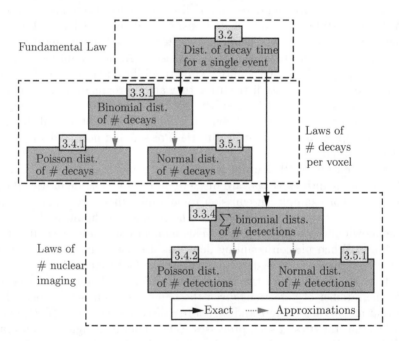

FIGURE 3.1 The summary of laws (models) used for statistical inferences (dist. = distribution). The numbers in the light boxes indicate section numbers where the laws are derived and described in details. The solid black arrows and gray dotted lines indicate the exact and approximate derivations, respectively.

3.2 FUNDAMENTAL STATISTICAL LAW

It is assumed that the imaging volume consists of a set of non-overlapping voxels. Each voxel of the image contains some number of radioactive nuclei that during the imaging time may undergo nuclear processes which are accompanied by a direct or indirect emission of photons. Figure 3.2 schematically presents a voxel with six radioactive nuclei contained. Although it seems that by assuming voxels the discrete image representation is postulated, the theory presented here is general and applies when the voxel size goes to zero and therefore applies in the limit to continuous representations.

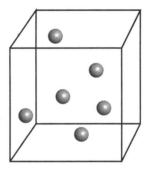

FIGURE 3.2 Schematic representation of a voxel with six radioactive nuclei shown as balls inside the voxel cube.

Let's consider a single nucleus. The chance dP that a nucleus decays at time t in some infinitesimal dt is proportional to dt regardless of the value of t assuming that the nucleus has not decayed before time t. We will use symbol λ to denote the proportionality between dP and dt. The λ is in units of second^{-1} and is referred to as the *decay constant*. The value of λ will be different for different radioactive nuclei but typically in nuclear imaging the same radioactive nuclei are used in single imaging session and therefore λ is the same for all nuclei under consideration. Suppose t_1 indicates the time at which the observation begins (the nucleus is in an excited state "ready" to decay). To mathematically express the chance that the nucleus has not decayed after t_1 and before time t, the time between t_1 and t is divided into small equal intervals dt as shown in Figure 3.3. If by $T = t - t_1$ we denote the duration of the observation, the number of those intervals is T/dt.

The chance that the decay has not occurred between t_1 and t is equal to the product of chances that it did not occur in any of the intervals dt (Figure 3.3) after t_1 and before t. Since λdt is the chance that it did occur, the $(1 - \lambda dt)$ indicates the probability that the decay did not occur in a single interval dt. Furthermore suppose that we would like to find the probability density $p_s(t; \lambda, t_1)$ for a single nucleus that the decay occurs at time t after we begin

FIGURE 3.3 The line represents the time from t_1 to t. The time is divided into small intervals dt.

the observation at time t_1. The decay constant and t_1 are known quantities (KQs). Using this definition of the p_s, the chance that the decay occurs during some short time dt at time t is therefore $p_s(t; \lambda, t_1)dt$.

Taking all this into consideration this can be summarized mathematically as the RHS of the following equation

$$\lim_{dt \to 0} p_s(t; \lambda, t_1)dt = \lim_{dt \to 0} (1 - \lambda dt)^{T/dt} \lambda dt \qquad (3.1)$$

and because $\lim_{x \to 0}(1 - ax)^{1/x} = e^{-a}$ for any number $a < \infty$ we obtain $p_s(t; \lambda, t_1)$ as $p_s(t; \lambda, t_1) = \lambda e^{-\lambda T}$. We see that the distribution depends only on the difference between t and t_1 and therefore without losing generality we assume the experiment is always started at $t_1 = 0$ which implies $T = t$. Using this we obtain

$$p_s(t; \lambda) = \lambda e^{-\lambda t}. \qquad (3.2)$$

This exponential relation reflects the fact that the longer the time t, the smaller the probability density of the decay at t because the nucleus under consideration had "more" chances to decay before that time.

Suppose that we consider an experiment with duration T. We implicitly assume that the observation (imaging) started at time 0. We introduce the new QoI b which is 1 when nucleus decays during the time T and it is 0 when it does not. It follows that

$$p(b = 1; \lambda, T) = \int_0^T p_s(t; \lambda)dt = 1 - e^{-\lambda T}. \qquad (3.3)$$

Since the RHS is dependent on the product of λ and T $p(b = 1; \lambda, T) = p(b = 1; \lambda T)$.

Reminder: As we stated in Chapter 1 we do not differentiate between probability density and probability mass which is evident in Equation (3.3) whereby the same symbol p and p_s we indicate the probability mass and probability density and reader should be mindful of this convention.

We state the *fundamental statistical law* used in this work as

$$p(b; \lambda T) = \begin{cases} 1 - e^{-\lambda T} & \text{for} \quad b = 1 \\ e^{-\lambda T} & \text{for} \quad b = 0 \end{cases}. \qquad (3.4)$$

The above is the probability that a single nucleus decays ($b = 1$) or not ($b = 0$) during time T. As stated above the law involves only a single binary QoI b, but all other statistical distributions used in PLD imaging will be derived from Equation (3.4).

3.3 GENERAL MODELS OF PHOTON-LIMITED DATA

In this section the binomial and multinomial distributions that describe the statistics of nucleus decay as well as the statistics of detections of counts are discussed.

3.3.1 BINOMIAL STATISTICS OF NUCLEAR DECAY

We consider a sample of radioactive material which contains r identical nuclei each with the same decay constant λ. For an easier transition to imaging later in the chapter, let's assume that the sample is contained within some voxel. In Figure 3.2, an example of such voxel is presented with six nuclei, but typically there will be many more. In general, the value of r will be unknown and considered the UQ. The QoI $\mathbf{b}$ which is a vector of r elements defines whether the nuclei in the voxel decayed or not in time T (elements of $\mathbf{b}$ are 1 or 0).

From the physics of nuclear decay it is known that the decays are independent. This means that whether one nucleus decays or not does not affect decays of others. Using this fact and Equation (3.4) the following joint distribution is obtained

$$p(\mathbf{b}|r; \lambda T) = \prod_{i=1}^{r} p(b_i; \lambda T) = (1 - e^{-\lambda T})^d (e^{-\lambda T})^{r-d}. \tag{3.5}$$

where $d \triangleq \sum_{i=1}^{r} b_i$ indicates the number of nuclei that decayed and $r - d$ indicates the number of nuclei that did not decay. For all $\mathbf{b}$s that have the same number of decays, d the probability of such $\mathbf{b}$ is the same as is evident in Equation (3.5). If now d is considered a QoI, then it follows that the probability that d number of decays occurred is the sum of all probabilities of $\mathbf{b}$s with the same d.

$$p(d|r; \lambda T) = \binom{r}{d}(1 - e^{-\lambda T})^d (e^{-\lambda T})^{r-d} \tag{3.6}$$

where $\binom{r}{d} = \frac{r!}{d!(r-d)!}$ is the number of different vectors $\mathbf{b}$ that have the same d and refer to as *binomial coefficient* for r and d. The above equation describes the *binomial distribution*. The binomial distribution describes Bernoulli process in which we have a number of independent identical "trials" where each trial can be either "success" (1) or "failure" (0). In those independent trials there is the same chance chance of the "success" of the trial which is another way of saying that trials are identically distributed or identical. It is evident that the Bernoulli process accurately describes the process of nuclear decay of single nucleus. The number of trials corresponds to the number of nuclei in the voxel. During the imaging time, each of the nuclei "tries" to decay and has a fixed chance of "success" described by the fundamental law described by Equation (3.4). The binomial distribution describes the probability of obtaining some number of successes. The number of successes is indicated as d and number of trials as r.

> **Notation:** In the following sections and chapters the notation would be quite convoluted if the dependence on known quantities (KQs) is indicated. In order to simplify the notation we will no longer indicate the explicit dependence on the KQs. Therefore we define $p(d|r; \lambda T) \triangleq p(d|r)$. We will conform to this convention of ignoring implicit dependence on KQs in notation throughout this book from now on. Occasionally exceptions from this rule will be made if lack of explicit dependence would introduce confusion or ambiguity.
>
> To simplify the notation and make it cleaner we also define $q \triangleq 1 - e^{\lambda T}$ to indicate the chance that a single nucleus decays during time T and with the decay constant λ. Finally, since the statistics of counting is defined based on discrete and non-negative numbers we will always assume that unless stated otherwise all QoIs are non-negative and all distributions are zero if any of the QoI is negative. This will simplify the notation considerably.

Using this new notation, the equation governing the statistics of decay of r nuclei is described by

$$p(d|r) = \binom{r}{d} q^d (1 - q)^{r-d}. \tag{3.7}$$

with d being the number of decays. The above conditional is defined for $d \le r$. We extend the definition to $d > r$ for which case $p(d|r) = 0$.

3.3.2   MULTINOMIAL STATISTICS OF DETECTION

Suppose that we have a measuring device that can detect photons emitted from a voxel as the result of the radioactive decay. We assume that the chance of a photon (or photons) emitted from the voxel of being registered by the measuring device is known. We indicate this chance by α which is constant and KQ. Since only a single detector is used in this example α can be considered as the sensitivity of the detection. If r is the total number of nuclei in the voxel as defined before, the number of detections y is

$$p(y|r) = \binom{r}{y} (\alpha q)^y (1 - \alpha q)^{r-y}. \tag{3.8}$$

It was assumed that a chance that the nucleus decays (q) and the chance that the resulting photon (or photons) are detected (α) are independent and therefore the chance that a decay occurs and is detected is αq.

Convention: We will frequently use the words "decay is detected." This is an abbreviation of the following "photons that resulted from nuclear decay are detected." Sometimes we will also use the term "count" to indicate detection of a decay. This convention is frequently utilized in the nuclear imaging field because of the nature of the detection process in which counts are registered (each count corresponds to a single detection). The term "event" is similar to the count but refers to decays that occur in imaging objects.

A very interesting property of the binomial distribution will be demonstrated below: the number of detections y and the total number of nuclei in a measured sample r are conditionally independent given the number of events d (number of decays). This means that if we know the number of events (decays) d, the knowledge of number of original nuclei r does not affect our beliefs about the number of detections y. Metaphorically speaking, the knowledge of d overshadows the effect of knowledge of r on beliefs about y.

The term $(1-\alpha q)^{r-y} = [(1-\alpha)q + (1-q)]^{r-y}$ in Equation (3.8) can be rewritten to $\sum_{x=0}^{r-y} \binom{r-y}{x}[(1-\alpha)q]^x (1-q)^{r-y-x}$ and Equation (3.8) transforms to:

$$p(y|r) = \sum_{x=0}^{r-y} \frac{r!}{y!(r-y)!} \frac{(r-y)!}{x!(r-y-x)!}(\alpha q)^y [(1-\alpha)q]^x (1-q)^{r-y-x} \quad (3.9)$$

Setting $x = d - y$, canceling terms $(r-y)!$ and adding $d!$ in the numerator and denominator we obtain

$$p(y|r) = \sum_{d=y}^{r} \frac{d!}{y!(d-y)!}\alpha^y (1-\alpha)^{d-y} \frac{r!}{d!(r-d)!}q^d(1-q)^{r-d} \quad (3.10)$$

In the above equation the right side $\frac{r!}{d!(r-d)!}q^d(1-q)^{r-d}$ is $p(d|r)$ using Equation (3.8) and therefore

$$p(y|r) = \sum_{d=y}^{\infty} \frac{d!}{y!(d-y)!}\alpha^y (1-\alpha)^{d-y} p(d|r). \quad (3.11)$$

On the other hand if the following identity is considered

$$p(y|r) = \sum_{d=0}^{\infty} p(y|d,r)p(d|r) = \sum_{d=y}^{\infty} p(y|d,r)p(d|r). \quad (3.12)$$

where the second equality comes from $p(y|d,r) = 0$ for $d < y$ by comparing it with Equation (3.11) we obtain that

$$p(y|d,r) = \frac{d!}{y!(d-y)!}\alpha^y (1-\alpha)^{d-y}. \quad (3.13)$$

Since the right-hand side is independent of r we arrive to

$$p(y|d,r) = p(y|d) = \frac{d!}{y!(d-y)!}\alpha^y(1-\alpha)^{d-y} \qquad (3.14)$$

which implies that y and r are conditionally independent given d (see Section 1.9). It is also clear that $p(y|d)$ has the binomial distribution.

Although at this point conditional independence of the measurement on the original number of nuclei may seem like a trivial advantage, it will become important when multi-dimensional conditionals are considered in subsequent sections. The importance of the conditional independence of the measurement and the actual number of radioactive nuclei will become evident in numerical calculations as well.

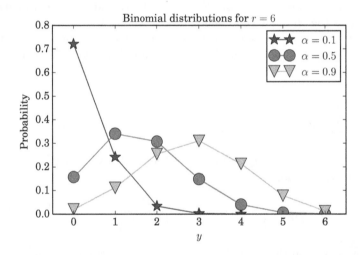

FIGURE 3.4 The discrete binomial distribution of the number of detections from a voxel containing $r = 6$ radioactive nuclei. Three types of symbols correspond to different detector sensitivities equal to $\alpha = 0.1$, $\alpha = 0.5$, and $\alpha = 0.9$. Data is represented by symbols (21 in total) and lines are added for improved clarity.

Example 3.1: Detection of ^{18}F decays from a voxel

Suppose we consider a voxel as pictured in Figure 3.2 with six radioactive nuclei of ^{18}F. Therefore, the value of r is known and equal to 6. We know that the half-life of ^{18}F is equal to $t_{1/2} = 109.77$ minutes. The observation time T is 120 minutes and sensitivity of the detector is $\alpha \in [0, 1]$. The goal is to find the distribution $p(y|r)$.

We find the λ using the value $t_{1/2}$. From the definition, the half-life is the time after which the probability of the decay $q = 1/2$. Therefore solving equation $q_{1/2} = (1 - e^{-\lambda t_{1/2}}) = 1/2$ provides the value of $\lambda = 1.05 \times 10^{-4} \mathrm{s}^{-1}$. The $p(y|r)$ can be now readily calculated from Equation (3.11) and it is shown in Figure 3.4 for different values of sensitivity α.

Scanners used in tomography have a large number of detector elements and therefore an event that occurs in some voxel can be detected in any of the detector elements or it may not be detected at all. Typically the sensitivity of the scanners (chance that if a decay occurs it is detected) is only a few percent. Schematically this is shown in Figure 3.5 where a single voxel and four detector elements are depicted .

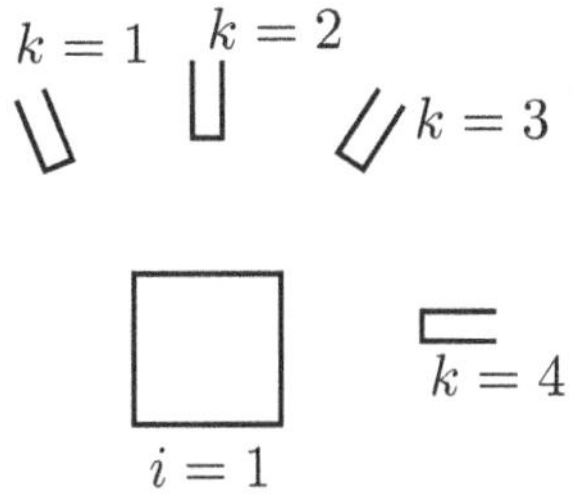

FIGURE 3.5 Schematic drawing of a single voxel with index $i = 1$ and multiple detector elements (four) indexed by $k = 1, \ldots, 4$.

Notation: Let $\mathbf{y}_1$ indicate a vector of the number of counts (detections of decays) that originated in a single voxel that is indicated by index 1. Although a single voxel is considered at this point and assigning an index to a single voxel seems unnecessary, it is done so the notation is consistent with the following sections in which many voxels are considered and indexing those voxels will be required. The elements of $\mathbf{y}_1$ are $(\mathbf{y}_1)_k$ where the index k is an index of the detector elements and runs from 1 to K. In other words $\mathbf{y}_1$ is the vector of size K. The α_{k1} indicates a chance of detection in detector element k of an event that occurred in voxel 1. Subscripts will also be used with QoIs related to a single voxel. For example d_1 indicates a number of decays in voxel 1 and r_1 indicates a total number of radioactive nuclei in voxel 1.

In Appendix C we derive from the basics the distribution $p(\mathbf{y}_1|d_1)$ (Equation (C.9)) of the number of detections in K detectors $\mathbf{y}_1$ if the number of emissions from the voxel d_1 is known. This distribution derived there is

$$p(\mathbf{y}_1|d_1) = \frac{d_1!}{\left(d_1 - \sum_{k'=1}^{K}(\mathbf{y}_1)_{k'}\right)! \prod_{k'=1}^{K}(\mathbf{y}_1)_{k'}!}$$

$$\times (1 - \epsilon_1)^{d_1 - \sum_{k'=1}^{K}(\mathbf{y}_1)_{k'}} \prod_{k=1}^{K}(\alpha_{k1})^{(\mathbf{y}_1)_k} \quad (3.15)$$

The above equation can be considerably simplified by introducing a new QoI c_1 which we define as the number of emissions from voxel 1 that was detected. Therefore, $d_1 \geq c_1$. Using this definition we define conditional

$$p(\mathbf{y}_1|c_1) = \begin{cases} \frac{c_1!}{\prod_{k'=1}^{K}(\mathbf{y}_1)_{k'}!} \prod_{k=1}^{K}\left(\frac{\alpha_{k1}}{\epsilon_1}\right)^{(\mathbf{y}_1)_k} & \text{for } \sum_{k=1}^{K}(\mathbf{y}_1)_k = c_1 \\ 0 & \text{for } \sum_{k=1}^{K}(\mathbf{y}_1)_k \neq c_1 \end{cases} \quad (3.16)$$

We also note that for a known d_1 the distribution of c_1 follows the binomial distribution

$$p(c_1|d_1) = \frac{d_1!}{c_1!(d_1 - c_1)!}(\epsilon_1)^{c_1}(1 - \epsilon_1)^{d_1 - c_1} \quad (3.17)$$

for $c_1 \leq d_1$ and zero otherwise.

Now, using Equations (3.16) and (3.17) the distribution of $p(\mathbf{y}_1|d_1)$ in Equation (3.15) transforms to a simple form

$$p(\mathbf{y}_1|d_1) = p(\mathbf{y}_1|c_1)p(c_1|d_1) \quad (3.18)$$

Significance of Equation (3.18): First, this equation implies conditional independence of $\mathbf{y}_1$ and d_1 given c_1. This fact will be stressed over and over again because it is one of the most important concepts used in this book.

Using a reasoning similar to the one that leads us to the derivation of $p(\mathbf{y}_1|d_1)$ (Equation (3.18)) demonstrated in Appendix C the $p(\mathbf{y}_1|r_1)$ can also be obtained. This yields the following distribution

$$p(\mathbf{y}_1|r_1) = \frac{c_1!}{\prod_{k'=1}^{K}(\mathbf{y}_1)_{k'}!} \prod_{k=1}^{K}\left(\frac{\alpha_{k1}}{\epsilon_1}\right)^{(\mathbf{y}_1)_k} \frac{r_1!}{c_1!(r_1 - c_1)!}(\epsilon_1 q)^{c_1}(1 - \epsilon_1 q)^{r_1 - c_1} =$$

$$= p(\mathbf{y}_1|c_1)p(c_1|r_1) \quad (3.19)$$

for $\sum_{k=1}^{K}(\mathbf{y}_1)_k = c_1$ and $c_1 \leq r_1$ and zero otherwise.

The distribution of the number of counts conditioned on the number of radioactive nuclei $p(\mathbf{y}_1|r_1)$ given above can be derived in an almost identical way as $p(\mathbf{y}_1|d_1)$ in Appendix C using αq instead of α.

Summary: In essence, if there is a known number of decays (d_1) or radioactive nuclei (r_1) in a single voxel and many detectors are used for registering counts, the distribution of those counts follows the multinomial distribution with the number of categories of this multinomial distribution equal to the number of detectors plus one. Depending on whether we consider d_1 or r_1 the K categories correspond to detections in K detectors and the additional category corresponds to (1) events that are not detected in any detector element or (2) nuclei that do not decay or if they did they are not detected in any detector element. We put much effort to derive these from the basic principles but the result is actually quite intuitive.

3.3.3 STATISTICS OF COMPLETE DATA

In the previous section the model of the number of counts detected which originated from a single voxel was derived. We extend the theory to multiple voxels. Since multiple voxels and multiple detector elements are used, the theory is general and can be considered as the **statistical model of nuclear imaging**. We present the illustrative figure of nuclear imaging scenario in Figure 3.6.

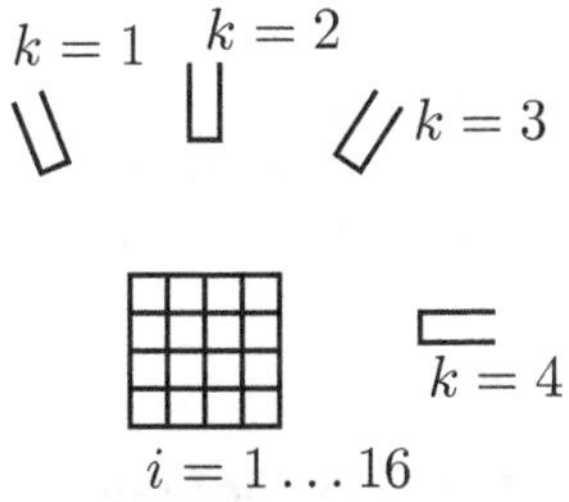

FIGURE 3.6 Schematic drawing of a multiple voxel image ($I = 16$) indexed by i and multiple detector elements ($K = 4$) indexed by k.

The results obtained in the previous section can be almost directly utilized by assuming the statistical independence of nuclear processes that occur in different voxels. In other words, decays in one voxel do not affect decays in other voxels. We will also assume that detections are independent, which indicates that a detection of a decay does not affect detections of other decays[1].

| **Important:** The independence of voxels and the independence of detections do not imply that the total numbers of detections (sums of detections

---

[1]This assumption may be violated by dead-time and pile-up effects in real scanners. We discuss these effects in Chapter 5.

from all voxels) in different detector elements are independent. In fact, numbers of detections are not independent which will be demonstrated later in this section.

We assume that the imaged volume consists of I voxels indexed by i and there are K detector elements indexed by k (Figure 3.6). We use the vector QoIs $\mathbf{y}_1 \ \mathbf{y}_2, \ldots, \mathbf{y}_I$ that indicate the number of events that occurred in voxels $1, 2, \ldots, I$ during a period of time T and were detected in detector elements. The length of each vector $\mathbf{y}_i$ is K.

Notation: The QoI $\mathbf{y}$ comprises all vectors $\mathbf{y}_1 \ \mathbf{y}_2, \ldots, \mathbf{y}_I$. The most straightforward way of thinking about $\mathbf{y}$ is to consider it as a vector of size $K \times I$ with elements y_{ki}. Each element of this vector is discrete (number of counts). The element y_{ki} indicates the number of events that occurs in voxel i and is detected in detector element k. To make the notation clear from this point on, we do not use brackets in $(\mathbf{y}_i)_k$ to indicate that we refer to kth element of vector $\mathbf{y}_i$ and instead use y_{ki} which refers to an element of vector $\mathbf{y}$ indicated by a pair of indices $\{k, i\}$. The QoI described by $\mathbf{y}$ is referred to as *complete data*. A *set* of all possible $\mathbf{y}$s is defined as $\mathbf{Y}$. In general, a set is a collection of objects (in our case, discrete vectors $\mathbf{y}$). We provide the basis of the set theory in Appendix B.

We define a matrix $\hat{\alpha}$ with elements α_{ki} as the *system matrix*. The elements α_{ki} describe the chance that if an event occurs in voxel i it will be detected in the detector element k. This value is a KQ and therefore the values are known at BE and AE stages.

Example 3.2: Simple Tomographic System (STS) (1) — Definition

A simple imaging system model is defined to illustrate many ideas presented in this chapter and in this book. The model is referred to as the *simple tomographic system* (STS). The STS consists of three voxels and three detector elements. Therefore $I = K$ and is equal to 3. Although the STS appears to be a trivial example, it will serve well to demonstrate many theoretical and computational concepts. It contains all statistical aspects of the real system. An interesting feature of the STS is that it allows a comparison of solutions obtained by the numerical methods with exact solutions obtained by using the algebra. It will also serve as an excellent illustration of several non-trivial constructs used in this book for which it would be much harder to gain an intuition when larger imaging systems are considered. The geometry of the system is presented in Figure 3.7. It is a tomographic system in which detector elements acquire data from exactly two pixels as demonstrated in Figure 3.7. Although we concentrate in this book on emission tomography, the system presented in Figure 3.7 can be used to investigate the transmission tomography systems as well.

The system matrix $\hat{\alpha}$ for this system is a 3×3 matrix:

$$\hat{\alpha} = \begin{bmatrix} \alpha_{11} & \alpha_{12} & 0 \\ \alpha_{21} & 0 & \alpha_{23} \\ 0 & \alpha_{32} & \alpha_{33} \end{bmatrix}. \tag{3.20}$$

Typically the system matrix in real imaging systems will be sparse with many of the elements equal to 0. This is reflected in the STS system matrix by including some zero valued elements. Obviously, having three elements equal to zero in 3×3 matrix hardly makes it sparse but it is our compromise in order to find a balance between the simplicity and the amount of a similarity with a real nuclear imaging system.

The form of matrix $\hat{\alpha}$ implies that the corresponding elements of $\mathbf{y}$ will be zero as well (see Equation (3.21)). If the chance that photon emitted in a voxel and detected in some detector element is zero, then the number of events emitted from that voxel and detected in the detector element must be zero as well. Therefore if the vector $\mathbf{y}$ is represented as a matrix with a similar layout as the system matrix the elements of $\mathbf{y}$ corresponding to zero-valued elements of the system matrix will also be zero. It follows that the general form of the complete data $\mathbf{y}$ for the system matrix specified in Equation (3.20) is

$$\mathbf{y} \equiv \left\{ \begin{bmatrix} y_{11} \\ y_{21} \\ 0 \end{bmatrix}, \begin{bmatrix} y_{12} \\ 0 \\ y_{32} \end{bmatrix}, \begin{bmatrix} 0 \\ y_{23} \\ y_{33} \end{bmatrix} \right\} \equiv \begin{bmatrix} y_{11} \\ y_{21} \\ 0 \\ y_{12} \\ 0 \\ y_{32} \\ 0 \\ y_{23} \\ y_{33} \end{bmatrix} \equiv \begin{bmatrix} y_{11} & y_{12} & 0 \\ y_{21} & 0 & y_{23} \\ 0 & y_{32} & y_{33} \end{bmatrix}. \tag{3.21}$$

The vectors $\mathbf{r}$, $\mathbf{d}$, and $\mathbf{c}$ all of size I describe the total number of radioactive nuclei per voxel (at the beginning of the scan), the number of nuclei that decayed per voxel, and the number of nuclei per voxel that decayed which was detected by the camera in any detector, respectively. All three QoIs are unobservable quantities.

Goal of imaging: The goal of any imaging study is to gain statistical inference about QoIs $\mathbf{r}$ or $\mathbf{d}$. These quantities are directly related to concentration of radioactive nuclei in the object. For example, the value of $\mathbf{r}$ directly indicates the concentration of the tracer in each voxel which is the typical endpoint in any nuclear imaging study.

Elements of those vectors r_i, d_i, and c_i specify the quantities for ith voxel of the image. Following our convention, we ignore in notation the dependence of the above QoIs on the *decay constant* λ and the time of the observation T (imaging time).

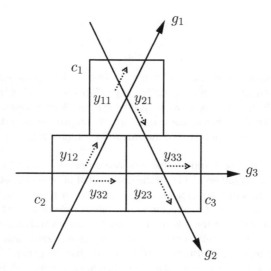

FIGURE 3.7 Simple tomographic system (STS). The g_1, g_2, and g_3 indicate the three detector elements. The voxels are indicated by c_1, c_2, and c_3, which correspond to the numbers of events that occurred in voxels and were detected. Non-zero values of $\mathbf{y}$ are also specified.

Since the decays in voxels and detection of counts that result from the decays are assumed independent, all distributions developed for a single voxel in the previous section (Section 3.3.1) can be extended to multi-voxel case just by multiplying distributions of each voxel. For example, using the independence, the distribution $p(\mathbf{y}|\mathbf{c})$ is obtained by

$$p(\mathbf{y}|\mathbf{c}) = p(\mathbf{y}|c_1,\ldots,c_I) = \frac{p(\mathbf{y}_1,\ldots,\mathbf{y}_I,c_1,\ldots,c_I)}{p(c_1,\ldots,c_I)} =$$

$$= \frac{p(\mathbf{y}_1,c_1)p(\mathbf{y}_2,c_2)\ldots p(\mathbf{y}_I,c_I)}{p(c_1)\ldots p(c_I)} = p(\mathbf{y}_1|c_1)p(\mathbf{y}_2|c_2)\ldots p(\mathbf{y}_I|c_I).$$

Using Equation (3.16) we obtain

$$p(\mathbf{y}|\mathbf{c}) = \prod_{i=1}^{I} \frac{c_i!}{\prod_{k=1}^{K} y_{ki}!} \prod_{k=1}^{K} \left(\frac{\alpha_{ki}}{\epsilon_i}\right)^{y_{ki}} = \prod_{i=1}^{I} \frac{c_i!}{(\epsilon_i)^{c_i}} \prod_{k=1}^{K} \frac{(\alpha_{ki})^{y_{ki}}}{y_{ki}!} \qquad (3.22)$$

The $p(\mathbf{y}|\mathbf{c})$ is interpreted as the probability of $\mathbf{y}$ (the complete data) for given $\mathbf{c}$. It follows that if $\exists \mathbf{y}_i : \sum_{k=1}^{K} y_{ki} \neq c_i$ the $p(\mathbf{y}|\mathbf{c}) = 0$. This definition constitutes the generalization of Equation (3.16) to many dimensions and explicitly given by

$$p(\mathbf{y}|\mathbf{c}) = \begin{cases} \prod_{i=1}^{I} \frac{c_i!}{(\epsilon_i)^{c_i}} \prod_{k=1}^{K} \frac{(\alpha_{ki})^{y_{ki}}}{y_{ki}!} & \text{if } \forall_i c_i = \sum_{k=1}^{K} y_{ki} \\ 0 & \text{if } \exists_i c_i \neq \sum_{k=1}^{K} y_{ki}. \end{cases} \quad (3.23)$$

Notation: If by $\mathbf{Y}$ we indicate a set of all possible $\mathbf{y}$s and define a subset $\mathbf{Y_c}$ such that $\mathbf{Y_c} = \{\mathbf{y} : \forall i \sum_{k=1}^{K} y_{ki} = c_i\}$. It follows that $\mathbf{Y_{c_1}} \cup \mathbf{Y_{c_2}} = \emptyset$ for $\mathbf{c}_1 \neq \mathbf{c}_2$. In other words $\mathbf{Y}$ is partitioned (see definition of set partition in Appendix B) into subsets $\mathbf{Y_c}$.

Example 3.3: Example of set partition

Suppose the experiment consists of simultaneously rolling two dice. The result is described by a two-dimensional vector $\mathbf{y}$ with the first coordinate describing the number on die 1 and the second the number on die 2. We define a QoI c which is the sum of the two numbers.

Subsets $\mathbf{Y_c}$ partition $\mathbf{Y}$ The set of possible pairs $\{y_1, y_2\} \in \mathbf{Y}$ (there are a total of 36 possibilities) is divided into 11 disjoint subsets $\mathbf{Y_c}$. Each subset corresponds to $c = 2, 3, \ldots, 12$. For example for $c = 4$ the $\mathbf{Y}_4 = \{1, 3\}, \{2, 2\}, \{3, 1\}$.

Probability $p(\mathbf{y}|c)$ Since obtaining any pair of numbers on two dice has the same chance of occurring, the $p(\mathbf{y}|c)$ can readily be calculated based on the size of each subset. For example for $c = 4$ for which the size of the subset is 3 the $p(\{1, 3\}|4) = p(\{2, 2\}|4) = p(\{3, 1\}|4) = 1/3$.

Using a similar approach that leads to derivation of $p(\mathbf{y}|\mathbf{c})$ other distributions $p(\mathbf{y}|\mathbf{d})$ and $p(\mathbf{y}|\mathbf{r})$ can be obtained and all derived distributions are summarized below. All three distributions are products of multinomial and binomial distributions.

MODELS OF COMPLETE DATA

Conditioned on the total number of detected decays

$$p(\mathbf{y}|\mathbf{c}) = \prod_{i=1}^{I} p(\mathbf{y}_i|c_i) = \begin{cases} \prod_{i=1}^{I} \frac{c_i!}{(\epsilon_i)^{c_i}} \prod_{k=1}^{K} \frac{(\alpha_{ki})^{y_{ki}}}{y_{ki}!} & \text{for } \mathbf{y} \in \mathbf{Y_c} \\ 0 & \text{for } \mathbf{y} \notin \mathbf{Y_c} \end{cases} \quad (3.24)$$

Conditioned on the number of decays ($y \in Y_c$ and $c \leq d^1$)

$$p(\mathbf{y}|\mathbf{d}) = \prod_{i=1}^{I} \frac{c_i!}{(\epsilon_i)^{c_i}} \left(\prod_{k=1}^{K} \frac{(\alpha_{ki})^{y_{ki}}}{y_{ki}!} \right) \frac{d_i!}{c_i!(d_i - c_i)!} (\epsilon_i)^{c_i} (1 - \epsilon_i)^{d_i - c_i} =$$

$$= \prod_{i=1}^{I} p(\mathbf{y}_i|d_i) = \prod_{i=1}^{I} p(\mathbf{y}_i|c_i)p(c_i|d_i) = p(\mathbf{y}|\mathbf{c})p(\mathbf{c}|\mathbf{d}) \quad (3.25)$$

^{1} $\mathbf{c} \leq \mathbf{d} \equiv \forall i : c_i \leq d_i$.

Conditioned on the number of radioactive nuclei ($y \in Y_c$ and $c \leq r^1$)

$$p(\mathbf{y}|\mathbf{r}) = \prod_{i=1}^{I} \frac{c_i!}{(\epsilon_i)^{c_i}} \left(\prod_{k=1}^{K} \frac{(\alpha_{ki})^{y_{ki}}}{y_{ki}!} \right) \frac{r_i!}{c_i!(r_i - c_i)!} (\epsilon_i q)^{c_i} (1 - \epsilon_i q)^{r_i - c_i} =$$

$$= \prod_{i=1}^{I} p(\mathbf{y}_i|r_i) = \prod_{i=1}^{I} p(\mathbf{y}_i|c_i)p(c_i|r_i) = p(\mathbf{y}|\mathbf{c})p(\mathbf{c}|\mathbf{r}) \quad (3.26)$$

^{1} $\mathbf{c} \leq \mathbf{r} \equiv \forall i : c_i \leq r_i$.

If the conditions given in the parentheses in headers of each conditional Equations (3.24)–(3.26) are not met, the distributions are zero. The above equations are the basic models for the statistical analysis used in this book. At this point these distributions do not relate measurements and the QoI. These relationships will be derived in the next section.

Conditional independence: The above equations imply that the $\mathbf{r}$ and $\mathbf{y}$ are conditionally independent given $\mathbf{c}$. To demonstrate this the definition of conditional independence given in Chapter 1 will be used. The $\mathbf{r}$ and $\mathbf{y}$ are conditionally independent given $\mathbf{c}$ if and only if $p(\mathbf{y}|\mathbf{r}, \mathbf{c}) = p(\mathbf{y}|\mathbf{c})$. This can be formally shown as follows:

$$p(\mathbf{y}|\mathbf{r}) = \sum_{\mathbf{c}'} p(\mathbf{y}, \mathbf{c}'|\mathbf{r}) = p(\mathbf{y}, \mathbf{c}|\mathbf{r}) \quad (3.27)$$

where we used the fact that from all possible $\mathbf{c}'$ only $p(\mathbf{y}, \mathbf{c})$ is non-zero for $c_i = \sum_{k=1}^{K} y_{ki}$. Continuing the proof

$$p(\mathbf{y}, \mathbf{c}|\mathbf{r}) = \frac{p(\mathbf{y}, \mathbf{c}, \mathbf{r})}{p(\mathbf{r})} = \frac{p(\mathbf{y}|\mathbf{c}, \mathbf{r})p(\mathbf{c}, \mathbf{r})}{p(\mathbf{r})} = p(\mathbf{y}|\mathbf{c}, \mathbf{r})p(\mathbf{c}|\mathbf{r}) \quad (3.28)$$

Comparing the RHS with Equation (3.26) we obtain

$$p(\mathbf{y}|\mathbf{c}, \mathbf{r}) = p(\mathbf{y}|\mathbf{c}) \quad (3.29)$$

that proves that conditional independence. Similarly it can be shown that
d and **y** are conditionally independent given **c**.

Explaining this in more intuitive terms, it can be said that if **c** (number
of decays per voxels that were detected) is known, the knowledge of **r** (total
number of radioactive nuclei in each voxel) is superfluous. The knowledge
of true value of **r** does not bring any additional information about **y** if
c is also known. Conditional independence is equivalent to saying that if
the number of decays that were detected in either detector is known, the
probability of number of detections in each detector is indifferent to the
number of decays (detected or not) and to the total number of radioactive
nuclei in the voxel.

Although the distribution of **y** conditioned on three different QoIs was
specified, it is clear that $p(\mathbf{y}|\mathbf{c})$ is the most important distribution among these
three. The other distributions ($p(\mathbf{y}|\mathbf{r})$ and $p(\mathbf{y}|\mathbf{d})$) can be obtained simply by
multiplying the $p(\mathbf{y}|\mathbf{c})$ by the appropriate product of binomial distributions
which is much easier to calculate for real imaging systems as it does not
involve the system matrix α. Essentially, based on the above, we will strive
in our approaches to divide the computational problem first into determining
some insights about **c** and then follow with analysis **d** or **r** if the analysis of
d or **r** is needed (see Chapter 6). Since the QoI **c** will be used over and over
again in this book the term *emission counts* (EC) is coined to refer to this
quantity.

Now let's consider a more elaborate example. In the following we demon-
strate how to calculate the $p(\mathbf{y}|\mathbf{c})$ for the STS system defined in Example 3.2.

**Example 3.4: Simple Tomographic System (STS) (2) — Model of com-
plete data**

In this example we will calculate the distribution values of the $p(\mathbf{y}|\mathbf{c})$ expressed
by Equation (3.24). In order to do so, the system matrix is specified as

$$\hat{\alpha} = \begin{bmatrix} \alpha_{11} & \alpha_{12} & 0 \\ \alpha_{21} & 0 & \alpha_{23} \\ 0 & \alpha_{32} & \alpha_{33} \end{bmatrix} = \begin{bmatrix} 0.05 & 0.02 & 0 \\ 0.15 & 0 & 0.1 \\ 0 & 0.18 & 0.1 \end{bmatrix}. \tag{3.30}$$

The sensitivity ϵ can be readily calculated from the definition $\epsilon_i = \sum_{k=1}^{K} \alpha_{ki}$ and
$\epsilon_1 = \epsilon_2 = \epsilon_3 = 0.2$ (the sum over the elements in each column). Therefore, for
this particular example the system sensitivity for detection of decays is uniform
and equal to 20% for each voxel.

Suppose that the numbers of decays per voxel **c** that were detected are
known and equal to $[5, 1, 1]$. The values of the model $p(\mathbf{y}|\mathbf{c})$ are calculated
in Table 3.1 using Equation (3.24). Although there are only seven detections
($\sum_{k=1}^{K} g_k = 7$), there are 24 different **y**s as it is apparent in Table 3.1 (Figure 3.8
shows the STS for this example). For real data sets the number of possible **y**s
is astronomical and it is impossible to actually consider all possible **y**s one by

one. On the right side of the table we provide the total number of detections at three detector elements that is implied by the values of $\mathbf{y}$ using $g_k = \sum_{i=1}^{I} y_{ki}$. The actual measurement made by the real imaging system will correspond to those numbers.

Examples of values of $p(\mathbf{y}|\mathbf{d})$ and $p(\mathbf{y}|\mathbf{r})$ are not provided because the number of possible $\mathbf{y}$s for the example vector $\mathbf{r} = [5, 1, 1]$ or $\mathbf{d} = [5, 1, 1]$ would be 896—a much too large number for an efficient presentation in the table. This large number comes from the fact that the sum of all detections $\sum_{i=1}^{I} \sum_{k=1}^{K} y_{ki}$ no longer has to be equal to 7 and can be lower which multiplies the number of cases that would have to be considered.

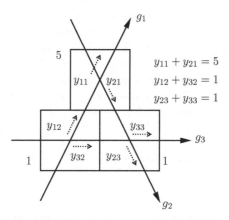

FIGURE 3.8 Schematics of the simple tomographic system (STS) as used in Example 3.4. The g_1, g_2, and g_3 indicate the three detector elements. The voxels 1, 2, and 3 emitted 5, 1, and 1 detected counts ($\mathbf{c}$).

3.3.4 POISSON-MULTINOMIAL DISTRIBUTION OF NUCLEAR DATA

So far, the distributions of the complete data $\mathbf{y}$ were established given by Equations (3.24)–(3.26). However, the complete data $\mathbf{y}$ cannot be observed directly (it is UQ) and therefore just by using the complete data we will not be able to obtain the model of the observable data $\mathbf{g}$. However, if we know the statistics of the complete data derived in the previous section, the statistics of $\mathbf{g}$ can be obtained simply by summing the distributions of complete data. In other words, the distribution of counts emitted from single voxel is multinomial and therefore the distribution of counts from all voxels is the sum of multinomial distributions from individual voxels. The sum of such distributions is governed by the Poisson-multinomial distribution which from

TABLE 3.1

Probability distribution $p(\mathbf{y}|\mathbf{c})$ calculated using Equation (3.24) for values of $\hat{\alpha}$ and c specified in Example 3.4 for the STS

#	\multicolumn{6}{c}{The value of non-zero^a elements y_{ki}}	$p(\mathbf{y}	\mathbf{c})$	\multicolumn{3}{c}{$g_k = \sum_{i=1}^{I} y_{ki}$}
	y_{11}	y_{21}	y_{12}	y_{32}	y_{23}	y_{33}		g_1	g_2	g_3
1	0	5	0	1	0	1	0.1068	0	5	2
2	1	4	0	1	0	1	0.1780	1	4	2
3	2	3	0	1	0	1	0.1187	2	3	2
4	3	2	0	1	0	1	0.0396	3	2	2
5	4	1	0	1	0	1	0.0066	4	1	2
6	5	0	0	1	0	1	0.0004	5	0	2
7	0	5	1	0	0	1	0.0119	1	5	1
8	1	4	1	0	0	1	0.0198	2	4	1
9	2	3	1	0	0	1	0.0132	3	3	1
10	3	2	1	0	0	1	0.0044	4	2	1
11	4	1	1	0	0	1	0.0007	5	1	1
12	5	0	1	0	0	1	0.0000	6	0	1
13	0	5	0	1	1	0	0.1068	0	6	1
14	1	4	0	1	1	0	0.1780	1	5	1
15	2	3	0	1	1	0	0.1186	2	4	1
16	3	2	0	1	1	0	0.0396	3	3	1
17	4	1	0	1	1	0	0.0066	4	2	1
18	5	0	0	1	1	0	0.0004	5	1	1
19	0	5	1	0	1	0	0.0119	1	6	0
20	1	4	1	0	1	0	0.0198	2	5	0
21	2	3	1	0	1	0	0.0132	3	4	0
22	3	2	1	0	1	0	0.0044	4	3	0
23	4	1	1	0	1	0	0.0007	5	2	0
24	5	0	1	0	1	0	0.0000	6	1	0

$$\sum = 1.0000$$

Note: The values of $p(\mathbf{y}|\mathbf{c})$ are given with the precision of four decimal places.
^{a} From the definition of the STS system $y_{31} = y_{22} = y_{13} = 0$. See Example 3.2.

definition described the sum of non-identically distributed multinomial distributions.

The Poisson-multinomial distribution is complex from the computational point of view and evaluation of likelihood based on this model is unpractical. Instead, for computational purposes we will use (Chapter 6) the model of the complete data $\mathbf{y}$ derived in the previous section to circumvent the complexity.

We follow with the formal specification of the Poisson-multinomial model in nuclear imaging. As before, the QoIs that we are interested in are $\mathbf{c}$, $\mathbf{d}$, and $\mathbf{r}$ depending on the needs for a task that is performed. Therefore, we will derive models of the data $p(\mathbf{g}|\mathbf{c})$, $p(\mathbf{g}|\mathbf{d})$, and $p(\mathbf{g}|\mathbf{r})$.

From the definition, an element of the data vector g_k is equal to $\sum_{i=1}^{I} y_{ki}$. By this definition we define a subset $\mathbf{Y_g}$ of the set $\mathbf{Y}$. We define $\mathbf{Y_g} = \{\mathbf{y} : \forall k \sum_{i=1}^{I} y_{ki} = g_k\}$. This definition is similar to the definition of $\mathbf{Y_c}$ provided in the previous section and illustrated by Example 3.3. Each of the subspaces $\mathbf{Y_g}$ is disjoint from each other $\mathbf{Y_{g_1}} \cup \mathbf{Y_{g_2}} = \emptyset$ for $\mathbf{g_1} \neq \mathbf{g_2}$. We previously defined a subset $\mathbf{Y_c}$ which was defined as $\mathbf{Y_c} = \{\mathbf{y} : \forall i \sum_{k=1}^{K} y_{ki} = c_i\}$. We now define an intersection between $\mathbf{Y_c}$ and $\mathbf{Y_g}$ as

$$\mathbf{Y_c} \cap \mathbf{Y_g} = \{\mathbf{y} : \forall i \sum_{k=1}^{K} y_{ki} = c_i, \forall k \sum_{i=1}^{I} y_{ki} = g_k\}. \tag{3.31}$$

Therefore, from the definition of the intersection $\mathbf{Y_c} \cap \mathbf{Y_g}$ we have

$$p(\mathbf{g}, \mathbf{c}|\mathbf{y}) = \begin{cases} 1 & \text{if } \mathbf{y} \in \mathbf{Y_c} \cap \mathbf{Y_g} \\ 0 & \text{if } \mathbf{y} \notin \mathbf{Y_c} \cap \mathbf{Y_g}. \end{cases} \tag{3.32}$$

We are ready now to derive the equation for the model of the data $p(\mathbf{g}|\mathbf{c})$

$$p(\mathbf{g}|\mathbf{c}) = \sum_{\mathbf{y} \in \mathbf{Y}} p(\mathbf{g}, \mathbf{y}|\mathbf{c}) = \sum_{\mathbf{y} \in \mathbf{Y}} \frac{p(\mathbf{g}, \mathbf{y}, \mathbf{c})}{p(\mathbf{c})} = \sum_{\mathbf{y} \in \mathbf{Y}} \frac{p(\mathbf{g}, \mathbf{c}|\mathbf{y})p(\mathbf{y})}{p(\mathbf{c})}$$

$$= \sum_{\mathbf{y} \in \mathbf{Y_c} \cap \mathbf{Y_g}} \frac{p(\mathbf{y})}{p(\mathbf{c})}$$

Because the sum is done over elements from subset $\mathbf{Y_c}$ for which $p(\mathbf{c}|\mathbf{y})$ is 1, we multiply the numerator by $p(\mathbf{c}|\mathbf{y})$ and use Bayes Theorem:

$$p(\mathbf{g}|\mathbf{c}) = \sum_{\mathbf{y} \in \mathbf{Y_c} \cap \mathbf{Y_g}} \frac{p(\mathbf{y})}{p(\mathbf{c})} = \sum_{\mathbf{y} \in \mathbf{Y_c} \cap \mathbf{Y_g}} \frac{p(\mathbf{c}|\mathbf{y})p(\mathbf{y})}{p(\mathbf{c})} = \sum_{\mathbf{y} \in \mathbf{Y_c} \cap \mathbf{Y_g}} p(\mathbf{y}|\mathbf{c}).$$

Therefore, the probability of data $\mathbf{g}$ if the number of decays that were detected $\mathbf{c}$ is known is provided by the following equation:

$$p(\mathbf{g}|\mathbf{c}) = \sum_{\mathbf{y} \in \mathbf{Y_c} \cap \mathbf{Y_g}} p(\mathbf{y}|\mathbf{c}). \tag{3.33}$$

Using Equation (3.23) we arrive to an important equation which provides the model of the imaging data given the number of counts per voxel that were detected. This equation will be a key for obtaining other models of the data given a number of radiative nuclei $\mathbf{r}$ or a number of decays $\mathbf{d}$.

$$p(\mathbf{g}|\mathbf{c}) = \sum_{\mathbf{y} \in \mathbf{Y_c} \cap \mathbf{Y_g}} \prod_{i=1}^{I} \frac{c_i!}{(\epsilon_i)^{c_i}} \prod_{k=1}^{K} \frac{(\alpha_{ki})^{y_{ki}}}{y_{ki}!} \tag{3.34}$$

Let's now consider computation of $p(\mathbf{g}|\mathbf{r})$ which is a probability of obtaining data $\mathbf{g}$ conditioned on the number of radioactive nuclei in voxels described by $\mathbf{r}$. We proceed similarly as in the case of the derivation of $p(\mathbf{g}|\mathbf{c})$ by defining the subset $\mathbf{Y_r}$ of complete data $\mathbf{Y}$ for which the $p(\mathbf{y}|\mathbf{r}) > 0$. Using Equation (3.26) this subset can be defined as a union of subsets $\mathbf{Y_c}$ for which $p(\mathbf{c}|\mathbf{r}) > 0$ and $\mathbf{c} \leq \mathbf{r}$. To formalize this using a mathematical equation we define

$$\mathbf{Y_r} = \bigcup_{p(\mathbf{c}|\mathbf{r})>0, \mathbf{c}\leq\mathbf{r}} \mathbf{Y_c}. \tag{3.35}$$

With that in mind starting with identity

$$p(\mathbf{g}|\mathbf{r}) = \sum_{\mathbf{c}} p(\mathbf{g}|\mathbf{c}, \mathbf{r})p(\mathbf{c}|\mathbf{r}), \tag{3.36}$$

and because of the conditional independence of $\mathbf{g}$ and $\mathbf{r}$ for given $\mathbf{c}$ ($p(\mathbf{g}|\mathbf{c},\mathbf{r}) = p(\mathbf{g}|\mathbf{c})$) and using Equation (3.33) we obtain

$$p(\mathbf{g}|\mathbf{r}) = \sum_{\mathbf{c}} \sum_{\mathbf{y}\in\mathbf{Y_c}\cap\mathbf{Y_g}} p(\mathbf{y}|\mathbf{c})p(\mathbf{c}|\mathbf{r}) = \sum_{\mathbf{y}\in\mathbf{Y_r}\cap\mathbf{Y_g}} p(\mathbf{y}|\mathbf{c})p(\mathbf{c}|\mathbf{r}). \tag{3.37}$$

Using a similar approach that leads to derivation of Equation (3.37), it can be shown that

$$p(\mathbf{g}|\mathbf{d}) = \sum_{\mathbf{y}\in\mathbf{Y_d}\cap\mathbf{Y_g}} p(\mathbf{y}|\mathbf{c})p(\mathbf{c}|\mathbf{d}). \tag{3.38}$$

The following example illustrates calculation of some of the probabilities of the observed data $\mathbf{g}$ derived above.

Example 3.5: Simple Imaging System (STS) (3) — Model of observed data

Let's use a particular setup of STS from Example 3.4 to illustrate concepts introduced in this section. We will use Table 3.1 where probabilities $p(\mathbf{y}|\mathbf{c})$ are calculated for $\mathbf{c} = [5, 1, 1]$. We will consider two examples of $\mathbf{g}$ for which we will find $p(\mathbf{g}|\mathbf{c})$. Suppose we are interested in obtaining the probability of obtaining $\mathbf{g} = [0, 4, 3]$. First, it has been already shown that set $\mathbf{Y_c}$ has 24 elements listed in Table 3.1. Because $p(\mathbf{g}|\mathbf{c})$ is the sum of probabilities over the intersection of $\mathbf{Y_c}\cap\mathbf{Y_g}$ out of these 24 cases we need to select the ones that also are members of $\mathbf{Y_g}$ (all 24 are already in $\mathbf{Y_c}$ set). The last three columns show gs for all ys in $\mathbf{Y_c}$. Since the vector $g = [0, 4, 3]$ is not there $\mathbf{Y_c} \cap \mathbf{Y_g} = \emptyset$ and therefore $p(\mathbf{g} = [0, 4, 3]|\mathbf{c} = [5, 1, 1]) = 0$.

For the second example let's consider determining the probability of $\mathbf{g} = [3, 3, 1]$ conditioned on $\mathbf{c} = [5, 1, 1]$. By examining Table 3.1 the $\mathbf{g} = [3, 3, 1]$ corresponds to row 9 and row 16. The intersection of $\mathbf{Y_c} \cap \mathbf{Y_g}$ is not empty and contains these two elements:

$$\mathbf{Y_{c=[5,1,1]}} \cap \mathbf{Y_{g=[3,3,1]}} = \left\{ \begin{bmatrix} 2 & 3 & 0 \\ 1 & 0 & 0 \\ 0 & 0 & 1 \end{bmatrix}, \begin{bmatrix} 2 & 3 & 0 \\ 1 & 0 & 0 \\ 0 & 1 & 0 \end{bmatrix} \right\}. \tag{3.39}$$

It follows that using Equation (3.33) or just by using entries in Table 3.1 corresponding to rows 9 and 16 we obtain the probability

$$p(\mathbf{g} = [3, 3, 1]|\mathbf{c} = [5, 1, 1]) = 0.0132 + 0.0396 = 0.0528 \qquad (3.40)$$

For the same reasons given in Example 3.4 we do not provide examples of values of $p(\mathbf{y}|\mathbf{d})$ and $p(\mathbf{y}|\mathbf{r})$ in the table because the sizes of subset $\mathbf{Y_r}$ and $\mathbf{Y_d}$ needed for calculation of those probabilities are too large to handle even for a meager total number of radioactive nuclei $\mathbf{r}$ or number of decays $\mathbf{d}$. Therefore the approach which involved consideration of all possibilities and direct calculations of those distributions from the definition is not practical even for the small toy STS considered here.

3.4   POISSON APPROXIMATION

In the medical imaging field the Poisson distribution is frequently used to describe the statistics of photon counting. The Poisson distribution is an approximation of the general statistics derived in previous sections. We demonstrate approximations that lead to Poisson distribution and investigate the accuracy of the approximations for emission tomography.

3.4.1   POISSON STATISTICS OF NUCLEAR DECAY

We examine first a case of the distribution that describes the number of emissions per voxel $p(d|r)$ given by Equation (3.7). This binomial distribution can be approximated by the *Poisson distribution* assuming $r \to \infty$ and defining a new QoI $f = r(1 - e^{-\lambda T}) \ll r$. The QoI f is interpreted as the expectation of the number of decays during time T with decay constant λ with r nuclei. This approximation is known as the *Poisson limit theorem*. The derivation of the Poisson distribution describing statistics of a number of decays is provided below. We start with the exact binomial distribution derived earlier and summarized by Equation (3.7).

$$p(d|r) = \binom{r}{d}(1 - e^{-\lambda T})^d (e^{-\lambda T})^{r-d} = \binom{r}{d}\left(\frac{f}{r}\right)^d \left(1 - \frac{f}{r}\right)^{r-d} =$$

$$= \binom{r}{d}\left(\frac{f}{r-f}\right)^d \left(1 - \frac{f}{r}\right)^r \qquad (3.41)$$

Now using $\lim_{x \to \infty}(1 - \frac{a}{x})^x = e^{-a}$ we obtain $\left(1 - \frac{f}{r}\right)^r \approx e^{-f}$. We also show that in the limit $\frac{1}{(r-f)^d}\binom{r}{d}$ can be approximated as follows

$$\lim_{r \to \infty} \frac{1}{(r-f)^d}\binom{r}{d} = \frac{1}{d!}\lim_{r \to \infty} \frac{r(r-1)\dots(r-d+1)}{(r-f)^d} \approx \frac{1}{d!}. \qquad (3.42)$$

Taking into account Equations (3.41) and (3.42), we obtain

$$p(d|r) \approx \frac{(qr)^d e^{-qr}}{d!} \tag{3.43}$$

Definition of activity: We define the *activity* f as $f \triangleq qr = r(1 - e^{-\lambda T})$. This is interpreted as the average (expected) number of decays during the observation period T for r nuclei with the decay constant λ. Note that the unit of f is described in our definition by the number of counts and not by the number of counts per unit of time.

It follows that

$$p(d|r) \approx \mathcal{P}(d|f) = \frac{(f)^d e^{-f}}{d!} \tag{3.44}$$

whereby $\mathcal{P}$ the Poisson distribution is indicated. The upper bound on error made in this approximation can be estimated using Le Cam's Theorem [64, 100] given next.

Le Cam's Theorem: Suppose we consider scalar independent QoIs $b_1, \ldots, b_r$ where each of bs can be either 0 or 1 with probability of being 1 equal to $p(b_\rho = 1) = p_\rho$ where $\rho \in \{1, \ldots, r\}$. There are a total of r QoIs b_ρ. We introduce a new QoI y which is $y = \sum_{\rho=1}^{r} b_\rho$. The distribution of such variable is known as *Poisson-binomial distribution* $p(y|p_1, \ldots, p_r)$ and does not have a closed-form expression. If $p_1 = \ldots = p_r = p$ the Poisson-binomial distribution becomes simple binomial distribution with r Bernoulli trials and chance of success p.

We define $x = \sum_{\rho=1}^{r} p_\rho$ and the Le Cam's Theorem in its original form states that:

$$\sum_{y=0}^{\infty} \left| p(y|p_1, \ldots, p_r) - \frac{x^y e^{-x}}{y!} \right| < 2 \sum_{\rho=1}^{r} p_\rho^2. \tag{3.45}$$

In other words the Poisson binomial distribution is approximated by the Poisson distribution with the maximum error of the approximation defined by Equation (3.45).

Using Le Cam's Theorem (Equation (3.45)), noting that for the case of nuclear decays the chance of success of a Bernoulli trial is equal for each trial, and $p_1 = \ldots = p_r = (1 - e^{-\lambda T})$

$$\sum_{d=0}^{\infty} |p(d|r) - \mathcal{P}(d|r, \lambda T)| < 2r(1 - e^{-\lambda T})^2 = \frac{2f^2}{r}. \tag{3.46}$$

From the above, it is clear that approximation works best if f is small compared to the total number of radioactive nuclei. Conversely, if the expected number of decays f is in the order of the total number of radioactive nuclei in the voxel, the approximation is poor. To demonstrate this, let's consider the following example.

Example 3.6: Decays of ^{201}Tl and ^{82}Rb from a voxel

The decay constant for ^{201}Tl is $\lambda_{Tl} = 2.64 \times 10^{-6} s^{-1}$. The ^{82}Rb half-life is 75.4 seconds and therefore $\lambda_{Rb} = 9.19 \times 10^{-3} s^{-1}$. Suppose we consider a voxel with $r = 10^9$ radioactive nuclei and $T = 5$ minutes. We consider two separate cases where the voxel contains 10^9 nuclei of ^{201}Tl and ^{82}Rb. Based on those data we can calculate the average number of decays for both isotopes and are equal to $f_{Tl} = 7.92 \times 10^5$ and $f_{Rb} = 9.37 \times 10^8$. Since f_{Rb} is comparable with $r = 10^9$ the Poisson approximation is poor and conversely since $f_{Tl} \ll r$ the Poisson approximation is good. This can be verified in Figure 3.9.

3.4.2 POISSON APPROXIMATION OF NUCLEAR DATA

The demonstration that statistics that governs the number of detections **g** (Poisson-multinomial distribution) can be approximated by the product of independent Poisson distributions requires a generalization of the Le Cam's Theorem [71] from binomial to multinomial distributions.

To arrive to this approximation consider a voxel and radioactive nuclei that it contains. One category of outcomes of what can happen to these nuclei is when they either do not decay or if they decay they are not detected. If this happens they are assigned to category 0 (Figure 3.10). If the nucleus decays and is detected in detector k it is assigned to category k.

If we consider the distribution of detections of decays occurring in a single voxel $p(\mathbf{y}_i|r_i)$ specified by Equation (3.26) distribution then after rearranging terms

$$p(\mathbf{y}_i|r_i) = \frac{r_i!}{(r_i - \sum_{k=1}^{K} y_{ki})! \prod_{k=1}^{K} y_{ki}!} (1 - \epsilon_i q)^{r_i - \sum_{k=1}^{K} y_{ki}} \prod_{k=1}^{K} (\alpha_{ki} q)^{y_{ki}}$$

$$= p(\mathbf{y}_i, y_{0i}|r_i) \quad (3.47)$$

whereby y_{0i} we indicate the number of nuclei that either did not decay or if they decayed were not detected (category 0). The equation defines a multinomial distribution with $K + 1$ categories with y_{ki} and $\alpha_{ki} q$ describing the number of counts and chance of obtaining a count in each of the 1 through K categories (detectors), respectively. Since the 0th category corresponds to a case for which the decay either did not occur or if it did it was not detected, the number of such events is denoted as y_{0i} and the chance that nucleus

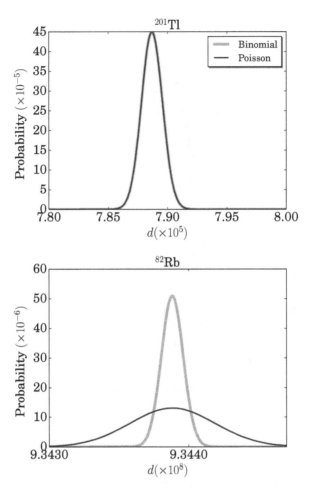

FIGURE 3.9 Comparison of binomial and Poisson distributions for ^{201}Tl (upper) and ^{82}Rb (lower).

falls within this category is $1 - \epsilon_i q$. We used $p(y_{0i}, \mathbf{y}_i | r_i) = p(\mathbf{y}_i | r_i)$ because the value of y_{0i} is implied by $\mathbf{y}_i$, and r_i $(y_{0i} = r_i - \sum_{k=1}^{K} y_{ki})$ therefore $p(y_{0i}, \mathbf{y}_i | r_i) = p(\mathbf{y}_i | r_i)$ that can be shown using the chain rule (Section 1.8.1).

From the definition, the number of acquired counts in detector elements $1, \ldots, K$ is described by a vector $\mathbf{g}$ which is a sum of contributions from each voxel with probability distributions of each of these contributions described by Equation (3.47). Since each of those contributions is distributed according to multinomial distributions which are not identically distributed, the sum of such distributions is described by the *Poisson-multinomial* distribution (see Appendix A for summary of various distributions including

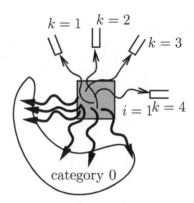

FIGURE 3.10 Schematic drawing of a single voxel (shaded) with index $i = 1$ and multiple detector elements (four) indexed by $k = 1, \ldots, 4$. The number of decays that were not detected and the number of nuclei that did not decay comprise category 0 of multinomial distribution.

Poisson-multinomial).

The Poisson–multinomial distribution of $\mathbf{g}$ can be approximated by the product of independent Poisson distributions using the generalization of Le Cam's Theorem [71].

> **Generalization of Le Cam's Theorem for multinomial distributions:** Suppose we consider I number of vector-QoIs $\mathbf{y}_i$ each of size K and elements y_{ki}. Each $\mathbf{y}_i$ is described by the multinomial distribution with $K + 1$ categories and probabilities of obtaining an outcome for each category p_{ki} with $k = 1, \ldots, K$ and $p_{0i} = 1 - \sum_{k=1}^{K} p_k$. Category 0 corresponds to an outcome which was not categorized in any of $1, \ldots, K$ other categories. The number of trials is r_i. In general the probabilities and the number of trials are different for each $\mathbf{y}_i$ and therefore the multinomial distributions are not identically distributed. If p_{0i}'s have high probabilities, the distribution of $p(g_1, \ldots, g_K)$ where $g_k = \sum_{i=1}^{K} y_{ki}$ can be approximated by the product of independent Poisson distributions $\mathcal{P}_k$ where the mean of $\mathcal{P}_k$ is $\mu_k = \sum_{i=1}^{I} p_{ki} r_i$. The bound on the error of this approximation of the Poisson-multinomial statistics with the product of independent Poisson distributions is
>
> $$\sum_{\mathbf{g} \in \mathbb{Z}^K} \left| p(\mathbf{g}|\mathbf{r}) - \prod_{k=1}^{K} \mathcal{P}(\mu_k) \right| \leq 2 \sum_{i=1}^{I} \left(\sum_{k=1}^{K} p_{ki} \right)^2 \qquad (3.48)$$

See McDonald [71] for the proof of the above inequality. It is evident that the above reduces to the Le Cam's bound (see the theorem on page 89) for $K = 1$ which corresponds to a case of binomial distribution.

Applying the theorem to nuclear imaging we have the means μ_k correspond to $\sum_{i=1}^{I} \alpha_{ki} q r_i$. Using the definition the voxel activity as $f_i = (1 - e^{-\lambda T}) r_i = q r_i$ we obtain the Poisson approximation as

$$p(\mathbf{g}|\mathbf{r}) \approx \prod_{k=1}^{K} \mathcal{P}(g_k|\mathbf{f}) = \prod_{k=1}^{K} \frac{(\sum_{i=1}^{I} \alpha_{ki} f_i)^{g_k} e^{\sum_{i=1}^{I} \alpha_{ki} f_i}}{g_k!}. \qquad (3.49)$$

The above equation is one of the most frequently used equations in medical imaging and defines the statistics governing the detection of discrete events with $\mathbf{f}$ interpreted as *voxel activity* which is defined as the expected number of decays in voxels during time T and with the decay constant λ. It has a desirable feature of conditional independence of elements of vector $\mathbf{g}$ given $\mathbf{f}$ making the analysis of the data nuclear data easier. The assumption of independence leads to a large number of algorithms developed in the past. For example, the *expectation maximization* algorithm used heavily for image reconstruction [62, 92] in nuclear imaging and countless other publications concerned with data analysis nuclear imaging use the assumption that data is Poisson distributed. For the general models, however, derived in previous sections ($p(\mathbf{g}|\mathbf{c})$, $p(\mathbf{g}|\mathbf{d})$, and $p(\mathbf{g}|\mathbf{r})$) the elements of $\mathbf{g}$ are not statistically independent (neither conditionally nor unconditionally).

Using the generalization of the Le Cam's Theorem the bound on the error (Equation (3.48)) can be specified as

$$2 \sum_{i=1}^{I} \left(\sum_{k=1}^{K} p_{ki} \right)^2 = 2 \sum_{i=1}^{I} \left(\sum_{k=1}^{K} q \alpha_{ki} r_i \right)^2 = 2 \sum_{i=1}^{I} f_i^2 \epsilon_i^2 \qquad (3.50)$$

where ϵ_i is the voxel sensitivity defined as $\epsilon_i = \sum_{k=1}^{K} \alpha_{ki}$.

3.5 NORMAL DISTRIBUTION APPROXIMATION

The normal (also sometimes referred to as Gaussian) distribution is one of the most important distributions in statistics. This is due to the fact that the *Central Limit Theorem* states that if a QoI is defined as a sum of a high number of some other QoIs, the distribution of the sum QoI follows the normal distribution with the assumption that QoIs that form the sum are independent and identically distributed. Sometimes the approximations used in the previous section and this section are named as the law of small numbers and law of large numbers for Poisson and normal distributions to indicate the conditions for which the approximations are accurate. However, this description is not completely accurate because for large numbers of counts both distributions Poisson and normal are good approximations of counting statistics as long as the mean of the distribution is much smaller than the original number of radioactive nuclei.

The normal distribution is of interest in medical imaging because counting statistics can be approximated by the normal distribution for some applications. Below, we provide the theory of approximations of binomial distribution that lead to the normal distribution.

3.5.1 APPROXIMATION OF BINOMIAL LAW

The binomial distribution is defined in Section 3.3.1 as the probability of d decays when mean number of radioactive nuclei is r and chance of decay is q. The Gaussian distribution is a reasonable approximation of binomial law for large number of decays (d). Note that in order for the Poisson approximation to be applicable $d \ll r$ and r is large and therefore the normal approximation seems less restrictive; however, it does not apply for small d for which Poisson approximation may be used if $d \ll r$.

To derive normal distribution we start from the general distribution of the number of decays for the total number of radioactive nuclei is r. This derivation follows Lesch and Jeske [68] and although it is only a heuristic approach it is intuitive and provides insights. The more formal derivation is provided next in Section 3.5.2. The binomial distribution is defined as

$$p(d|r) = \binom{r}{d} q^d (1 - q)^{r-d} \tag{3.51}$$

where $q = (1 - e^{-\lambda t})$. Suppose we consider a ratio of $p(d+1|r)/p(d|r)$ which is equal to

$$\frac{p(d+1|r)}{p(d|r)} = \frac{(r-d)q}{(d+1)(1-q)} \tag{3.52}$$

Here we introduce a new variable $x = (d - rq)/\sigma$ where $\sigma = \sqrt{q(1-q)r}$. We also introduce a new distribution $p^*(x) = p(x\sigma + rq|r)$. With this substitution the ratio $\frac{p(d+1|r)}{p(d|r)}$ becomes:

$$\frac{p^*(x + 1/\sigma)}{p^*(x)} = \frac{(r - x\sigma - qr)q}{(x\sigma + qr + 1)(1 - q)} \tag{3.53}$$

By rearranging the terms and using $\sigma^2 = q(1-q)r$ we obtain

$$\frac{p^*(x + 1/\sigma)}{p^*(x)} = \frac{\sigma^2 - xq\sigma}{\sigma^2 + x(1-q)\sigma + (1-q)} = 1 - \frac{1 - xq/\sigma}{1 + x(1-q)/\sigma + (1-q)/\sigma^2} \tag{3.54}$$

Taking the logarithm of both sides and dividing both sides by $1/\sigma$

$$\frac{\log p^*(x + 1/\sigma) - \log p^*(x)}{1/\sigma}$$

$$= \frac{\log(1 - xq/\sigma) - \log(1 + x(1-q)/\sigma + (1-q)/\sigma^2)}{1/\sigma}$$

At this point the approximation is made assuming that $1/\sigma$ is small (which indicates that $q(1-q)r$ is large) the left-hand side is approximated by the derivative of $\log p^*(x)$ and the logarithms on the right-hand side are approximated as $\log(1 \pm \delta) \approx \pm\delta$. We obtain

$$\frac{d\log(p^*(x))}{dx} \approx x + (1-q)/\sigma \approx x \tag{3.55}$$

Solving the above yields

$$p^*(x) = Ce^{-x^2/2}. \tag{3.56}$$

The C is the integration constant. Now substituting to original QoI and finding C by normalizing we obtain:

$$p(d|r) \approx p^*(x) = \frac{1}{\sigma\sqrt{2\pi}}e^{\frac{(d-f)^2}{2\sigma^2}} \tag{3.57}$$

where $f = rq$ and $\sigma^2 = q(1-q)r$. The approximation is good for large values of $q(1-q)r$. Interestingly, contrary to popular beliefs, a large number of expected events qr is not sufficient to guarantee that normal approximation of binomial is accurate.

3.5.2 CENTRAL LIMIT THEOREM

The normal distribution of the number of decays per sample (shown in the previous section) as well as the approximation of the counting statistics in detector elements can be derived from the *central limit theorem* (CLT) [67] given below and multi-variate central limit theorem provided later in this section.

Central limit theorem: In the classical version of the theorem we assume that $\mathbf{y}_i = y_{1i}, y_{2i}, \ldots, y_{Ki}$ describe K QoIs each with the identical independent distribution (i.i.d) describing the uncertainty about the QoI. The mean and variance of distributions of each of ys is the same and equal to $\bar{y}_i$ and σ_i^2, respectively. The CLT states that the distribution of the $\sum_{k=1}^{K} y_{1i} + y_{2i} + \ldots + y_{Ki}$ approaches the normal distribution for $K \to \infty$. The mean of the limit normal distribution is $K\bar{y}_i$ and variance $K\sigma_i^2$.

Note: Typically the classical CLT is stated in terms of the distribution of the mean of the QoI (in our case we used the sum) but the statements are equivalent.

As described in the previous sections, the distribution of the number of decays d is described as the sum large number of Bernoulli trials where each with the mean of q and variance $q(1-q)$. Therefore by means of the CLT we approximate the distribution of d with the normal approximation the mean rq and variance $q(1-q)r$. This result is identical to the distribution derived in the previous section and summarized by Equation (3.57).

If we are interested in the approximation that leads to multi-dimensional distribution of the number of counts detected in detector elements the derivation is slightly more complex. First, let's state the multi-variate version of the CLT [67].

> **Multi-dimensional central limit theorem:** Suppose we consider I vector QoIs $y_1, y_2, \ldots, y_I$ with each y_i drawn from i.i.d. Let's by Σ indicate the $K \times K$ covariance matrix between each of the elements of vectors y_i. The covariance matrix is the same for each y_i per the assumption of the identical distribution. The mean of each y_i is denoted by $\bar{y}_i$ with elements $\bar{y}_{ki}$. The multidimensional CLT states that $g = \sum_{i=1}^{I} y_k$ follows multivariate normal distribution with means $\bar{g}$ where $\bar{g}_k = \sum_{i=1}^{I} \bar{y}_{ki}$ and covariance $\Sigma_g = I\Sigma$ expressed by the following:
>
> $$\mathcal{N}(g|\bar{g}, \Sigma_g) = \frac{1}{\sqrt{(2\pi)^K |\Sigma_g|}} e^{-\frac{1}{2}(g-\bar{g})^T \Sigma_g^{-1}(g-\bar{g})} \qquad (3.58)$$

The distribution of counts in detector element was shown to follow Poisson–multinomial distribution which is the sum of multinomial distributions with unequal means. Although distributions are independent due to independence of the number of decays in different voxels, they are not identically distributed as the covariance matrix is different for each of the multinomial distributions that form the sum. In general for each y_i there is a different covariance matrix Σ_i. It follows that the multivariate CLT [67] cannot be used directly as stated above because the assumption of identical distribution is violated.

Although the CLT cannot be used directly, suppose we consider a $\mathcal{N}(g|\bar{g}, \Sigma_{\bar{g}})$ where $\Sigma_{\bar{g}}$ is defined as follows:

$$\bar{\Sigma}_g = \sum_{i=1}^{I} \Sigma_i \qquad (3.59)$$

It can be shown [112] that the difference between the sum of independent non–identically distributed vector QoI and multivariate normal distribution defined by $\bar{\Sigma}_g$ is bounded and decreases with higher number of elements of the sum. We will use this to approximate the distribution of the number of detected counts with the multivariate normal distribution.

First, we consider a vector of means $\bar{y}_i$. Since the distribution of y_i is multinomial the elements of mean vector $\bar{y}_i$ are

$$\bar{y}_{ki} = \alpha_{ki} r_i q \qquad (3.60)$$

Using the property that the variance of element of a vector y_{ki} that is multinomially distributed is $\alpha_{ki}(1 - \alpha_{ki})r_i q$ and the covariance between the elements

k and k' is $-\alpha_{ki}\alpha_{k'i}r_i q^2$ (see Appendix A) we have that the covariance matrix Σ_i is

$$\Sigma_i = \begin{bmatrix} \alpha_{1i}(1 - q\alpha_{1i})r_i q & -\alpha_{1i}\alpha_{2i}r_i q^2 & \dots & -\alpha_{1i}\alpha_{ki}r_i q^2 \\ -\alpha_{2i}\alpha_{1i}r_i q^2 & \alpha_{2i}(1 - q\alpha_{2i})r_i q & \dots & -\alpha_{2i}\alpha_{ki}r_i q^2 \\ \dots & \dots & \dots & \dots \\ -\alpha_{ki}\alpha_{1i}r_i q^2 & -\alpha_{ki}\alpha_{2i}r_i q^2 & \dots & \alpha_{ki}(1 - q\alpha_{ki})r_i q \end{bmatrix} \quad (3.61)$$

Using results stated above we approximate the distribution of counts in detectors by multivariate normal with the mean equal to

$$\bar{g}_k = \sum_{i=1}^{K} \bar{y}_{ki} = \sum_{i=1}^{I} \alpha_{ki}r_i q = \sum_{i=1}^{I} \alpha_{ki}f_i \quad (3.62)$$

where f_i is the average number of emissions per voxel in duration of the imaging study (same definition as used before). The covariance (Equations (3.59) and (3.61)) is approximated as follows:

$$\bar{\Sigma}_{\mathbf{g}} = \sum_{i=1}^{I} \Sigma_i = \sum_{i=1}^{I} r_i q \begin{bmatrix} \alpha_{1i}(1 - q\alpha_{1i}) & -\alpha_{1i}\alpha_{2i}q & \dots & -\alpha_{1i}\alpha_{ki}q \\ -\alpha_{2i}\alpha_{1i}q & \alpha_{2i}(1 - q\alpha_{2i}) & \dots & -\alpha_{2i}\alpha_{ki}q \\ \dots & \dots & \dots & \dots \\ -\alpha_{ki}\alpha_{1i}q & -\alpha_{ki}\alpha_{1i}q & \dots & \alpha_{ki}(1 - q\alpha_{ki}) \end{bmatrix}$$

$$= \sum_{i=1}^{I} f_i \begin{bmatrix} \alpha_{1i}(1 - q\alpha_{1i}) & -\alpha_{1i}\alpha_{2i}q & \dots & -\alpha_{1i}\alpha_{ki}q \\ -\alpha_{2i}\alpha_{1i}q & \alpha_{2i}(1 - q\alpha_{2i}) & \dots & -\alpha_{2i}\alpha_{ki}q \\ \dots & \dots & \dots & \dots \\ -\alpha_{ki}\alpha_{1i}q & -\alpha_{ki}\alpha_{2i}q & \dots & \alpha_{ki}(1 - q\alpha_{ki}) \end{bmatrix} \quad (3.63)$$

The above approximation of statistics in detector elements using multivariate normal distribution predicts negative cross-covariance values between different detector elements, which is expected because more counts detected in one detector element indicates that less counts are available for detection in other detectors (the total number of possible detections is limited by the total number of nuclei). In the Poisson limit where q is small the covariance matrix reduces to the diagonal matrix. The diagonal elements are approximated as $\alpha_{ki}(1 - q\alpha_{ki}) \approx \alpha_{ki}$ and cross covariance terms $-\alpha_{k'i}\alpha_{ki}q \approx 0$. For such case

$$\bar{\Sigma}_{\mathbf{g}} \approx \begin{bmatrix} \sum_{i=1}^{I} \alpha_{1i}f_i & 0 & \dots & 0 \\ 0 & \sum_{i=1}^{I} \alpha_{2i}f_i & \dots & 0 \\ \dots & \dots & \dots & \dots \\ 0 & 0 & \dots & \sum_{i=1}^{I} \alpha_{ki}f_i \end{bmatrix} \quad (3.64)$$

The diagonal elements corresponds to the mean of counts at the detectors (see Equation (3.62)) which is a familiar result coming from Poisson distribution approximation.

4 Monte Carlo methods in posterior analysis

This section introduces *Monte Carlo (MC) methods*. MC methods are used [14, 85] in almost all areas of science and engineering. The Bayesian approaches greatly benefit from using MC methods combined with the high computational power of computers. This chapter is not meant as a thorough discussion of MC methods and a complete review can be found in Robert and Casella [85]. The goal of this chapter is to introduce methods that are used in medical imaging and are specific to nuclear imaging data analysis used in this book. In our applications the MC methods will be used for the calculation of approximations of posterior or marginalized posterior distributions called in general *posterior computing*.

In the first part of this chapter we introduce posterior computing for contiguous and discrete quantities. At first, it will be assumed that a perfect algorithm that provides independent samples from probability distributions is available. In most practical applications such algorithm will not be known and we will resort to *Markov Chain Monte Carlo* (MCMC) methods discussed in Section 4.4. The MCMC methods will be used for a generation of samples from large dimensional distributions of continuous and discrete quantities.

4.1 MONTE CARLO APPROXIMATIONS OF DISTRIBUTIONS

This section may seem paradoxical because the distributions will be estimated under assumption that samples from the same distributions can be obtained. It may seem that if samples are available, the distribution itself is known and there is a little sense of estimating it. In the large majority of problems however even if the distribution is known the marginalized distribution will not be easily available. The methods provided in this section serve as MC approaches to estimate marginalized distributions if sampling algorithms is available. We build distributions directly from the samples by histogramming the samples. For other approaches to MC estimations of distributions see Gelfand and Smith [29].

4.1.1 CONTINUOUS DISTRIBUTIONS

The methods discussed in this chapter will gently introduce the reader to posterior computing. Suppose we consider some vector of size I of contiguous QoIs $\mathbf{f}$ with each element i taking values from $-\infty$ to ∞ and that we are only interested in a single element of vector $\mathbf{f}$. Without losing generality, the element of interest is f_1, and therefore in order to determine the posterior of this quantity, the marginalization needs to be done

$$p(f_1) = \int_{-\infty}^{\infty} df_2 \ldots \int_{-\infty}^{\infty} df_I p(\mathbf{f}) \qquad (4.1)$$

Notation: For clarity of the presentation, we will use notation $p(f_1)$ and $p(\mathbf{f})$ to indicate 1– and I-dimensional posterior distributions without specifying a QoI that these distributions are conditioned on (e.g., $p(f_1|\mathbf{G} = \mathbf{g})$). It should be understood that all distributions that are used in this chapter are posteriors unless specified otherwise.

Equation (4.1) is an example of marginalizations that lead to a one-dimensional posterior which later can be used for statistical inference. However, the approach can be generalized to marginalizations of the multi-dimensional posteriors. For example, the joint posterior of first m elements of $\mathbf{f}$ is

$$p(f_1, \ldots, f_m) = \int_{-\infty}^{\infty} df_{m+1} \ldots \int_{-\infty}^{\infty} df_I p(\mathbf{f}) \qquad (4.2)$$

By specifying the first m elements the generality is not lost because the joint is insensitive to order of QoIs. Therefore, the elements can be ordered in such a way that QoIs to be marginalized have labels $m + 1, \ldots, I$.

Reminder: Following the convention adopted in this book we refer to probability mass and probability density as probability. Therefore although Equations (4.1) and (4.2) define probability density they are referred to as probability.

All tasks specified by Equations (4.1) and (4.2) may be difficult to accomplish when standard numerical integrations methods [40] are used. The MC computational approach to this problem is strikingly simple and effective. Suppose we are able to acquire samples from the distributions in such a way that the probability of getting a sample $p^*(\mathbf{f}^*)$ is proportional to the posterior $p(\mathbf{f} = \mathbf{f}^*)$ and R such samples are acquired. We use $\mathbf{f}^*$ to indicate that a particular value of QoI $\mathbf{f}$. The $p(\mathbf{f} = \mathbf{f}^*)$ indicates a single value of the posterior for $\mathbf{f} = \mathbf{f}^*$. For the problems defined by Equation (4.1) we define region $[f_1^* - \delta_1, f_1^* + \delta_1]$ and count the number of samples with values of f_1 that fall within this region. The number of those samples is $R_{f_1^*}$. When counting the samples, all other than f_1 coordinates of $\mathbf{f}$ are ignored and therefore whether the sample falls in region $[f_1^* - \delta_1, f_1^* + \delta_1]$ is decided based on a single coordinate. The MC approximations of the marginalized posterior value $\hat{p}(f_1^*)$ is

$$\hat{p}(f_1^*) = \frac{R_{f_1^*}}{2\delta_1 R}. \qquad (4.3)$$

When multidimensional marginalized density is required, the approach is similar. If a marginalized probability at $f_1^*, \ldots, f_m^*$ is needed and R samples are acquired the estimator of the marginalized probability is equal to

$$\hat{p}(f_1^*, f_2^*, \ldots, f_m^*) = \frac{R_{f_1^*, f_2^*, \ldots, f_m^*}}{2^m R \prod_{i=1}^{m} \delta_i} \tag{4.4}$$

where $R_{f_1^*, f_2^*, \ldots, f_m^*}$ is the number of samples that fell within the volume of m–dimensional box defined by $f_1^* \pm \delta_1, \ldots, f_m^* \pm \delta_m$.

Let's consider an error that is made by the above approximation. In the following, only the case of MC approximation of Equation (4.1) will be considered but similar analysis can be done to estimate the error of approximation of Equation (4.2). The error made during this approximation of Equation (4.1) comes from two sources. One source of the error is statistical in nature coming from the finite number of samples. The value of $R_{f_1^*}$ is obtained in a random process that is governed by the binomial distribution $\mathcal{B}(p_{\delta_1}; R)$ where p_{δ_1} is the probability that a first coordinate of the sample f^* falls within $[f_1^* - \delta_1, f_1^* + \delta_1]$ and

$$p_{\delta_1} = \int_{f_1^* - \delta_1^*}^{f_1^* + \delta_1^*} p(f_1) df_1. \tag{4.5}$$

Since the $R_{f_1^*}$ is sampled from the binomial distribution $\mathcal{B}(p_{\delta_1}; R)$ (see Appendix A), the variance of $R_{f_1^*}$ is $R p_{\delta_1}(1 - p_{\delta_1})$ and therefore the variance of $\hat{p}$ due to statistical error is

$$var(\hat{p}_{f_1^*}) = \frac{p_{\delta_1}(1 - p_{\delta_1})}{4 \delta_1^2 R} \tag{4.6}$$

where we abbreviated $\hat{p}(f_1 = f_1^*)$ to $\hat{p}_{f_1^*}$. From Equation (4.6) it is clear that a higher number of samples will result in a smaller variance. The variance is also dependent on the size of the region δ_1. If the region is relatively small, the p_{δ_1} will be approximately proportional to δ_1 and therefore the variance approximately proportional to the inverse of δ_1. This is in agreement with intuitive reasoning as with larger δ the number of samples falling within $[f_1^* - \delta_1, f_1^* + \delta_1]$ will be higher and statistical error should decrease.

Although it seems that by increasing δ_1 the variance will be reduced, there is the other source of error, the bias that will increase with larger size of δ_1. The source of bias is introduced due to the assumption that $\int_{f_1^* - \delta_1}^{f_1^* + \delta_1} p(f_1) df_1 = 2\delta_1 p(f_1^*)$. To find the approximation of the value of bias let's again assume that δ_1 is small and using Taylor expansion

$$p(f_1) \approx p(f_1 = f_1^*) + p'(f_1 = f_1^*|)f + \frac{1}{2}p''(f_1 = f_1^*)f^2 \tag{4.7}$$

and therefore the $\int_{f_1^* - \delta_1}^{f_1^* + \delta_1} p(f_1) df_1$ can be approximated using Equation (4.7) as

$$\int_{f_1^* - \delta_1}^{f_1^* + \delta_1} p(f_1) df_1 \approx 2\delta_1 p(f_1 = f_1^*) + \frac{1}{3}\delta_1^3 p''(f_1 = f_1^*). \tag{4.8}$$

It follows that bias of $\hat{p}$ is

$$bias(\hat{p}) \approx \frac{1}{3}\delta_1^3 p''(f_1 = f_1^*) \tag{4.9}$$

Since both sources of errors (variance and bias) are independent, the total error ε^2 is described by the following

$$\varepsilon^2(\hat{p}_{f_1^*}) = bias^2(\hat{p}_{f_1^*}) + var(\hat{p}_{f_1^*}) = \frac{1}{9}\delta_1^6[p''(f_1 = f_1^*)]^2 + \frac{p_{\delta_1}(1 - p_{\delta_1})}{4\delta_1^2 R} \tag{4.10}$$

It is clear that there is some optimal value for δ_1 for which the total error is minimal. Unfortunately in practical applications, the expression in Equation (4.10) can be evaluated only if $p''(f_1 = f_1^*)$ and p_{δ_1} are known. In practice the bias will usually be ignored as its contribution to the total error is typically much smaller than the error from the variance. Some examples of the applications of error analysis are provided in Example 4.1.

To illustrate the concepts consider the following example of the bivariate normal distribution. Although the theory provided so far and the example below demonstrate the MC marginalization to a single dimensional distribution, the extension to the multidimensional marginalized posterior is straightforward.

Example 4.1: Bivariate normal – sampling

Properties of bivariate normal distribution:
We choose to use as an example the multivariate normal distribution that has a general form

$$p(\mathbf{f}) = \frac{1}{\sqrt{(2\pi)^2|\Sigma|}}e^{-(\mathbf{f}-\boldsymbol{\mu})^T\Sigma^{-1}(\mathbf{f}-\boldsymbol{\mu})}. \tag{4.11}$$

For the bivariate case $\mathbf{f} = [f_1, f_2]$. The $\boldsymbol{\mu}$ is the vector of means equal to $[\mu_1, \mu_2]$ and Σ is the covariance equal to

$$\Sigma = \begin{bmatrix} \sigma_1^2 & \sigma_{12} \\ \sigma_{12} & \sigma_2^2 \end{bmatrix} \tag{4.12}$$

We normalize the off-diagonal elements of the covariance with respect to σ_1 and σ_2 as follows $\sigma_{12} = \nu\sigma_1\sigma_2$ where ν is chosen so the specified relation between σ_{12} and σ_1 and σ_2 holds. The inverse of the covariance matrix is

$$\Sigma^{-1} = \frac{1}{1-\nu^2}\begin{bmatrix} 1/\sigma_1^2 & -\nu/(\sigma_1\sigma_2) \\ -\nu/(\sigma_1\sigma_2) & 1/\sigma_2^2 \end{bmatrix}. \tag{4.13}$$

The bivariate normal distribution is then

$$p(\mathbf{f}) = \frac{1}{2\pi\sigma_1\sigma_2\sqrt{1-\nu^2}}e^{-\frac{1}{2(1-\nu^2)}\left[\left(\frac{f_1-\mu_1}{\sigma_1}\right)^2 + 2\nu\left(\frac{f_1-\mu_1}{\sigma_1}\right)\left(\frac{f_2-\mu_2}{\sigma_2}\right) + \left(\frac{f_2-\mu_2}{\sigma_2}\right)^2\right]}.$$

Having the joint distribution $p(\mathbf{f})$ the conditional $p(f_1|f_2)$ can be estimated using marginalization and chain rules introduced in Section 1.8.1. The values of the joint for f_2 are extracted and then normalized to 1 which yields

$$p(f_1|f_2) = \frac{1}{\sqrt{2\pi\sigma_1^2(1-\nu^2)}}e^{-\frac{1}{2}\frac{\left(f_1-\frac{\nu\sigma_1}{\sigma_2}(f_2-\mu_2)\right)^2}{(1-\nu^2)\sigma_1^2}} \tag{4.14}$$

The above is described as the normal distribution

$$\mathcal{N}\left(f_1; \frac{\nu\sigma_1}{\sigma_2}(f_2-\mu_2), \sqrt{(1-\nu^2)\sigma_1^2}\right). \tag{4.15}$$

If analytically marginalized over f_1 the joint $p(\mathbf{f})$ reduces to the standard normal distribution in 1D

$$p(f_2) = \frac{1}{\sqrt{2\pi\sigma_2^2}}e^{-\frac{1}{2}\left(\frac{f_2-\mu_2}{\sigma_2}\right)^2}. \tag{4.16}$$

The above equations provide the exact analytic equations that can be used to determine the true value of the densities. They will serve as the gold standard and compared with values estimated using MC methods.

Monte Carlo draws from bivariate normal:

Before drawing samples from the bivariate normal distribution, a method of drawing samples from a single-dimensional normal distribution needs to be established. One of the most frequently used algorithms to draw samples from normal distribution is based on the Box–Muller transform [12]. The algorithm consists of three steps and produces two independent draws from $\mathcal{N}(0,1)$.

1. Obtain two draws x_1 and x_2 from a uniform distribution $\mathcal{U}(0,1)$
2. Transform drawn values to $x_1' = \sqrt{-2\log(x_1)}$ and $x_2' = 2\pi x_2$
3. Obtain two independent normal draws x_1'' and x_2'' by $x_1'' = x_1'\cos(x_2')$ and $x_2'' = x_1'\sin(x_2')$

For the derivation of this algorithm refer to Box and Muller [12]. To obtain a sample x_1''' from $\mathcal{N}(\mu,\sigma)$ if a sample x_1'' from $\mathcal{N}(0,1)$ is available (e.g., obtained by the Box–Muller algorithm) the value needs to be scaled and shifted according by

$$x_1''' = \sigma x_1'' + \mu \tag{4.17}$$

The above illustrates the approach for drawing samples from a normal distribution. To obtain draws from the bivariate normal it is sufficient to draw f_2 from the normal described by Equation (4.16) and then to draw from the normal conditional Equation (4.14) for that value of f_2 obtained in the first draw. Figure 4.1 shows an example of 500 such draws from a bivariate normal with $\mu = [5,5]$ and $\Sigma = \begin{bmatrix} 1 & -0.5 \\ -0.5 & 2 \end{bmatrix}$. As it is evident from this example, obtaining draws from bivariate is relatively easy. Direct draws were possible and an algorithm for doing so was given in this example. However, in a large majority of real-world problems direct draws will not be possible. When direct draws are not possible methods such as Markov chains discussed in Section 4.4 can be used.

Total error and sample error:

The histogram shown in Figure 4.1 at the top can be used to estimate the values of marginal distribution of $p(f_1)$ at the centers of each bar of the histogram. This is done by dividing the number of samples obtained in each histogram bin by the total number of samples and the histogram width. Equation (4.10) provides an estimate of the total error of such estimation. Figure 4.2 schematically presents the extent of the total error. The bars are indicative of the square root of the total error as defined in Equation (4.10). Intuitively, if a Monte Carlo 500-sample run is performed, estimates of the values of densities f_1^*'s indicated by centers of histogram bars will be distributed approximately normally with a spread (standard deviation) indicated by the bars.

One such sample is shown in Figure 4.3. The error bars are the MC error estimated from the sample using Equation (4.18).

The total error of MC approximation of continuous distributions was calculated under the assumption of known value of p_{δ_1}. This however will seldom be the case as it requires knowledge of original distribution that we are trying to estimate. In order to obtain the approximation of the total error, we assume that bias component is much smaller than the variance and p_{δ_1} are much smaller than 1 so $(1 - \delta_1) \approx 1$. Using these two assumptions and approximating p_{δ_1} by its estimator, the total error becomes:

$$\varepsilon^2(\hat{p}_{f_1^*}) \approx \frac{R_{f_1^*}}{4\delta_1^2 R^2} \tag{4.18}$$

The above is the Monte Carlo error for estimation of marginalized densities estimated at f_1^* where $R_{f_1^*}$ is the number of samples obtained in the histogram bin with width $2\delta_1$ centered at f_1^*. The R is the total number of Monte Carlo samples.

4.1.2 DISCRETE DISTRIBUTIONS

Conceptually, discrete distributions are easier to use with MC marginalizations and approximations of distributions because the bias component of the error is zero and therefore the analysis of Monte Carlo errors becomes much easier. This is under assumption that we use a histogram for every discrete value of quantity for which the marginalization is sought.

Suppose a QoI $\mathbf{y}$ is considered and a set of all possible $\mathbf{y}$s is denoted as $\mathbf{Y}$, which we assume is equal to $\mathbb{Z}^K$ ($\mathbb{Z}$ indicates a set of integer numbers)

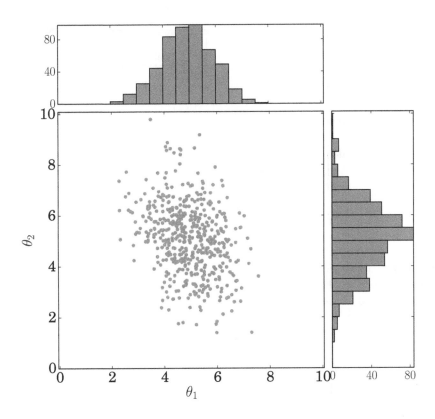

FIGURE 4.1 500 samples drawn from the bivariate normal distribution with $\mu = [5, 5]$ and Σ specified by values $1, -0.5, -0.5$, and 2 are shown in the center. Histograms are shown with bin width of 0.5 centered at values $0.5i$ where $i \in \mathbb{Z}$.

where K is the number of dimension of QoI $\mathbf{y}$. Furthermore, only the y_1 is of interest. The marginalized probability of $p(y_1)$ is

$$p(y_1) = \sum_{y_2=-\infty}^{\infty} \sum_{y_3=-\infty}^{\infty} , \ldots, \sum_{y_K=-\infty}^{\infty} p(\mathbf{y}) \qquad (4.19)$$

Suppose that R samples from the distribution $p(\mathbf{y})$ are available and we are interested in marginalized $p(y_1 = y_1^*)$. The probability of obtaining y_1^* in a single sample is $p_{y_1^*}$. Therefore the R samples can be considered as R Bernoulli trials and distribution of $R_{y_1^*}$ follows the binomial distribution. For such distribution the expectation of the number of samples for a particular y_1^* is $p(y_1 = y_1^*)R$ and therefore it is the unbiased estimator of the $p(y_1 = y_1^*)$ is

$$\hat{p}(y_1 = y_1^*) = \frac{R_{y_1^*}}{R}. \qquad (4.20)$$

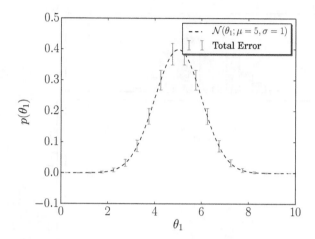

FIGURE 4.2 Total error from MC experiment indicated by bars. The line corresponds to a normal distribution (the marginal of the bivariate normal (Equation (4.16)) used in this example).

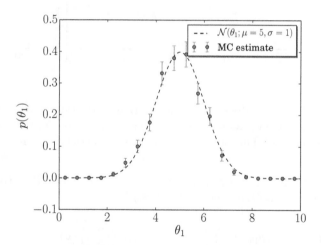

FIGURE 4.3 Monte Carlo estimates of the density values indicated by the points. The error bars correspond to the estimate of MC error using Equation (4.18). The line corresponds to a normal distribution (the marginal of the bivariate normal (Equation (4.16)) used in this example).

The variance of $R_{y_1^*}$ is equal to $Rp(y_1 = y_1^*)(1 - p(y_1 = y_1^*))$ since $R_{y_1^*}$ is binomially distributed. The variance of the estimator:

$$var(\hat{p}(y_1 = y_1^*)) = var(R_{y_1^*}/R) = p(y_1 = y_1^*)(1 - p(y_1 = y_1^*))/R. \quad (4.21)$$

Using the estimator of $\hat{p}(y_1 = y_1^*)$ the sample variance (calculated from the sample) is therefore

$$var\left(\hat{p}(y_1 = y_1^*)\right) \approx \frac{R_{y_1^*}(R - R_{y_1^*})}{R^3}. \tag{4.22}$$

Since typically $R_{y_1^*} \ll R$ we obtain

$$var(\hat{p}(y_1 = y_1^*)) \approx \frac{R_{y_1^*}}{R^2}. \tag{4.23}$$

Suppose that in a single MC simulation, there are several different posterior values of $p(y_1 = y_1^{*1}), p(y_1 = y_1^{*2}), \ldots, p(y_1 = y_1^{*m})$ that we need to estimate. Each MC sample can be considered as a Bernoulli trial with m categories. Therefore, the distribution of a number of samples for $R_{y_1^{*1}}, R_{y_1^{*2}}, \ldots, R_{y_1^{*m}}$ is distributed according to the multinomial distribution. It follows from the properties of multinomial distribution that the expectation and variance will be the same as in the case when only a single value of the posterior was needed in Equations (4.20) and (4.23). However, the covariance of the estimates will be non-zero equal to

$$cov\left(\hat{p}(y_1 = y_1^{*1}), \hat{p}(y_1 = y_1^{*2})\right) = \frac{p(y_1 = y_1^{*1})p(y_1 = y_1^{*2})}{R} \approx \frac{R_{y_1^{*1}}R_{y_1^{*2}}}{R^3} \tag{4.24}$$

and similarly for other pairs of estimators.

The ratio of covariance (Equation (4.24)) to variance (Equation (4.23)) is

$$\frac{cov\left(\hat{p}(y_1 = y_1^{*1}), \hat{p}(y_1 = y_1^{*2})\right)}{var(\hat{p}(y_1 = y_1^{*1}))} \approx \frac{R_{y_1^{*2}}}{R}. \tag{4.25}$$

Based on the above, the MC covariance can typically be ignored for $R_{y_1^{*1}} \ll R$ and $R_{y_1^{*2}} \ll R$.

4.2 MONTE CARLO INTEGRATIONS

One of the most useful applications of MC methods in Bayesian statistics is the numerical approximation of integrals. In general, the integrals

$$\Phi = \int_A^B \phi(f)p(f)df. \tag{4.26}$$

are considered. This is an integral $\int_A^B$ some function $\phi(.)$ calculated over the density $p(f)$. A and B indicate the integration limits. A single dimensional example is used but the extension to the multidimensional case is straightforward. As before, it is assumed that independent samples from $p(f)$ can be obtained and in the course of Monte Carlo run a total of R samples obtained.

Let's divide the $[A, B]$ into non-overlapping equal intervals Δ. It is assumed that there are K intervals and $B - A = K\Delta$. In this formulation each sample drawn from $p(f)$ can be considered as a Bernoulli trial. The outcome of this trial can fall within any of the K categories. The numbers of outcomes in intervals are therefore distributed according to a multinomial distribution. The number of categories is equal to the number of intervals, the number of Bernoulli trials is equal to the number of samples. The probability of obtaining outcome for a category k is equal to $p_k = \int_{\Delta_k} p(f)df$. With a total of R trials the number of outcomes in each category is R_k.

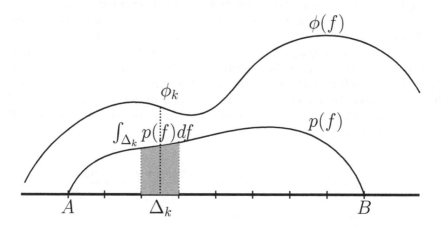

FIGURE 4.4 Monte Carlo sampling from interval $[A, B]$ with sampling density $p(f)$. The chance that a MC sample falls within the Δ_k is equal to $p_k = \int_{\Delta_k} p(f)df$ as illustrated.

Using the properties of multinomial distributions (Appendix A) the expectation of the number of outcomes R_k for the interval Δ_k is

$$E(R_k) = Rp_k = R \int_{\Delta_k} p(f)df \qquad (4.27)$$

the variance and covariance are

$$var(R_k) = Rp_k(1 - p_k) \text{ and } cov(R_k, R_{k'}) = Rp_k p_{k'} \qquad (4.28)$$

The MC estimator of the integral $\hat{\Phi}$ is defined as

$$\hat{\Phi} = \frac{1}{R} \sum_{k=1}^{K} \phi_k R_k \qquad (4.29)$$

where ϕ_k is the value of the integrated function at the center of interval Δ_k (see Figure 4.4). With $\Delta \to 0$ the expectation of this estimator $\hat{\Phi}$ is evaluated as follows

$$E(\hat{\Phi}) = E\left(\frac{1}{R}\sum_{k=1}^{K}\phi_k R_k\right) = \frac{1}{R}\sum_{k=1}^{K}\phi_k E(R_k) \overset{Equation\ (4.27)}{=} \sum_{k=1}^{K}\phi_k \int_{\Delta_k} p(f)df$$

$$\overset{\Delta_k \to 0}{=} \sum_{k=1}^{K}\int_{\Delta_k}\phi(f)p(f)df \overset{\Delta_k \to 0}{=} \int_A^B \phi(f)p(f)df. \quad (4.30)$$

The above demonstrates that the result of the Monte Carlo integration using estimator $\hat{\Phi}$ on average will yield the integral (it is unbiased). The variance of the estimator is calculated as follows

$$var(\hat{\Phi}) = var\left(\frac{1}{R}\sum_{k=1}^{K}\phi_k R_k\right)$$

$$= \frac{1}{R^2}\left(\sum_{k=1}^{K}var(\phi_k R_k) - \sum_{k=1}^{K}\sum_{k'=1, k'\neq k}^{K}cov(\phi_k R_k, \phi_{k'} R_{k'})\right)$$

$$\overset{Eq.Equation\ (4.28)}{=} \frac{1}{R^2}\left(\sum_{k=1}^{K}\phi_k^2 R p_k(1-p_k) - \sum_{k=1}^{K}\sum_{k'=1, k'\neq k}^{K}\phi_k\phi_{k'} R p_k p_{k'}\right)$$

$$= \frac{1}{R}\left(\sum_{k=1}^{K}\phi_k^2 p_k - \sum_{k=1}^{K}\phi_k p_k\sum_{k'=1}^{K}\phi_{k'} p_{k'}\right).$$

Using $p_k = \int_{\Delta_k} p(f)df$ and taking Δ_ks to 0 we obtain

$$var(\hat{\Phi}) = \frac{1}{R}\left[\int_A^B \phi^2(f)df - \left(\int_A^B \phi(f)df\right)^2\right] = \frac{\sigma_\phi^2}{R}. \quad (4.31)$$

where σ_ϕ^2 is $E^{p(f)}(\phi^2) - (E^{p(f)}(\phi))^2$. For $R \to \infty$ the variance goes to zero. This together with unbiasedness shows that with an infinite number of samples the estimate is equal to the integral. This property of MC integration is usually shown by applying the law of large numbers [26, 27] but we were able to demonstrate this by using properties of the multinomial distribution.

With $\Delta \to 0$ the value of R_k in Equation (4.29) will tend to be either 0 or 1 and therefore the equation is

$$\hat{\Phi} = \frac{1}{R}\sum_{j=1}^{R}\phi_j \quad (4.32)$$

Using Equation (4.31) and standard estimator of the standard deviation σ_ϕ the variance of this estimator is

$$\hat{var}(\hat{\Phi}) = \frac{1}{R-1}\sum_{j=1}^{R}\left(\phi_j - \hat{\Phi}\right)^2 \quad (4.33)$$

The above derivation can be performed for multidimensional integrals as well. An interesting property of MC estimator is that the variance is not dependent on dimensionality of the integration which can be shown if Equation (4.31) is derived for the multidimensional case (an exercise left for the reader).

4.3 MONTE CARLO SUMMATIONS

The following general form of the summation is considered. Only a simple summation is discussed in this section but as was the case for integration the extension of the methods developed here to multiple summations is straightforward. Suppose we consider a discrete QoI g defined on a subset of integers $\Omega = [A, B]$ and some function $\Phi(g)$. The sum to estimate is

$$\Phi = \sum_{g=A}^{B} \phi(g)p_g \tag{4.34}$$

where $\sum_{k=A}^{B} p_g = 1$. The estimate of the sum can be done in a very similar manner as in the case of integration over continuous quantity f in Section 4.2. It is assumed that samples can be drawn from discrete distribution of p_g and R of such samples are drawn. The Monte Carlo estimate of the sum $\hat{\Phi}$ is then

$$\hat{\Phi} = \frac{1}{R} \sum_{j=1}^{R} \phi_j \tag{4.35}$$

where ϕ_j is the value of ϕ for g drawn in kth draw. The estimate of variance of this estimate is

$$v\hat{a}r(\hat{\Phi}) = \frac{1}{R-1} \sum_{j=1}^{R} \left(\phi_j - \hat{\Phi}\right)^2 \tag{4.36}$$

Example 4.2: Numerical summation

We will use Example 1.7 in Chapter 1. In this example the joint distribution

$$p(f, g) = \frac{6}{(\pi g)^2} \frac{f^g e^{-f}}{g!} \tag{4.37}$$

where $f \in [0, \infty]$ and $g \in \mathbb{N}$ was marginalized to

$$p(f) = \frac{6e^{-f}}{\pi^2} \sum_{g=1}^{\infty} \frac{f^g}{g^2 g!} \tag{4.38}$$

Here, in this example the sum in the equation above will be evaluated using numerical methods. We note first that if we define $p^*(g) = 6/(\pi g)^2$ and be able to draw R samples from this distribution the density $p(f)$ can be approximated by

$$\hat{p}(f) = \frac{e^{-f}}{R} \sum_{k=1}^{R} \frac{f^{g_k}}{g_k!}$$ (4.39)

where g_k is the value of g for k–th sample drawn from $p(g) = 6/(\pi g)^2$.

An algorithm for drawing a sample from $p(g) = 6/(\pi g)^2$ is illustrated in Figure 4.5. This algorithm may not be optimal in terms of efficiency but illustrates well the principles in drawing the samples from discrete distributions. In this example the distribution is simple and direct draws are possible. For the actual computation used in this example the maximum number g that was used was 10^3. It was chosen arbitrarily to avoid numerical instabilities. Since the integrated function $\frac{f^{g_k}}{g_k!}$ is very small for large values of g the selection of maximum value of g does not significantly affect the result of the calculation.

When considered Equation (4.38) the samples can also be drawn from $p^*(g) = e^{-f} f^g / g!$ which is the Poisson distribution $\mathcal{P}(g|f)$. If R samples are drawn from the Poisson distribution then the summation can be estimated as

$$\hat{p}(f) = \frac{1}{R} \sum_{k=1}^{R} \frac{1}{g_k^2}$$ (4.40)

We expected that both approaches should provide the identical estimates which was demonstrated by results shown in Figure 4.6.

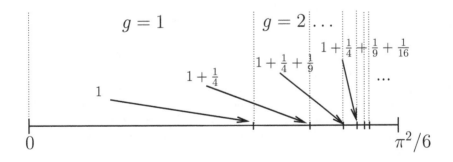

FIGURE 4.5 Generation of samples from $p(g) = 6/(\pi g)^2$. First, one sample x from $\mathcal{U}(0, \pi^2/6)$ is obtained. If the value of $x \leq 1$ $g = 1$ then sample is 1. Otherwise, if $1 < x \leq 1 + \frac{1}{4}$ sample is 2. Otherwise if $1 + \frac{1}{4} < x \leq 1 + \frac{1}{4} + \frac{1}{9}$ sample is 3, etc.

4.4 MARKOV CHAINS

In previous sections of this chapter we provided several examples of calculation of marginalizations, integrals, and sums under the assumption that it is

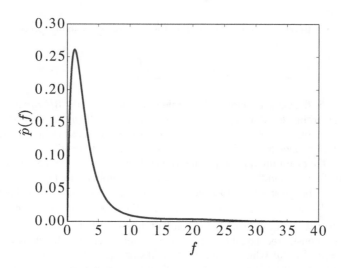

FIGURE 4.6 Estimates of values of $\hat{p}(f)$ obtained at 0.1 intervals (connected by line) using Equations (4.39) and (4.40) and $R = 10000$. Both estimates perfectly overlap.

possible to draw samples from probability distributions. More ofter than not there will be no clear way to design an algorithm for direct draws of samples. TO draw samples from distributions in this book we will rely on a class of methods based on *Markov chains*.

The Markov process is such that any prediction about the future depends only on the current *Markov state* of the process. In other words, the history of how the "state of nature" evolved to the current state is irrelevant for the prediction of the future evolution of the system. In this book we will use the term Markov chain to indicate a series of Markov states generated by the Markov process. In all cases considered, finite/countable chains, which are an ordered collection of Markov states generated by the Markov process, will be used. Continuous Markov chains [34, 93] (e.g., Wiener processes, Brownian motions, diffusion) can also be considered but they are not covered in this nuclear imaging book.

Example 4.3: Repetitive coin toss

Repetitive coins tossing is one of the easiest examples of a Markov process. We start by tossing a coin and counting 1 for heads (H) and -1 for tails (T). The total count defines the Markov state. For example if 10 consecutive tosses were $\{H, H, H, T, H, T, T, H, H, H\}$, then the Markov chain constructed using specified rules is $\{1, 2, 3, 2, 3, 2, 1, 2, 3, 4\}$. Although it seems that a 10th Markov state obtained by tossing H when current state was 3 depends on all the tosses that occurred before, it is actually irrelevant how the chain arrived to 3. It could have actually gotten there in millions of Markov moves. All we care about is that the current state is 3 and to generate the next state based solely on this value.

An example of a non-Markovian chain is a chain in which the next value in the chain depends on the last two states. For example it adds count 1 if the last two tosses are the same (HH,TT) and subtract 1 if they are different (HT, TH). For our example of $\{H, H, H, T, H, T, T, H, H, H\}$ the chain would be $\{U, 1, 2, 1, 0, -1, 0, -1, 0, 1\}$ where U stands for undefined.

Relevance of Markov chains to numerical integration/summation/marginalization: If properly designed, the Markov chains can represent a collection of samples from a given probability distribution.

The creation of the appropriate Markov chain with such property may be quite challenging. There are several issues that need to be addressed that are detailed in subsequent sections.

4.4.1 MARKOV PROCESSES

We introduce Markov processes using the concept of the *state space* [76]. We will assume that we are concerned with some *state space*[1] Ω which can be for example a domain over which the integration or summation is performed. In general, it is a space over which the samples are to be obtained. The state space can either be continuous or discrete and with any dimensions. The probability measure $p(s)$ is defined over the space (for example the $p(s)$ can be the posterior). Therefore, for each state in the state space a probability measure is assigned that obeys axioms introduced in Chapter 1. We define the Markov process over the state space such that if the system is in state s it generates state u. The process generates u in a random fashion, that is, it is not guaranteed that it will generate u every time the process is in state s. The idea of Markov processes is to design an algorithm that generates new states which will have some desirable properties explained later .

The probability (or probability density for continuous Ω) that u is generated if the system is in s is called *transition probability* $P(s \to u)$. For a well-conditioned Markov process all transition probabilities satisfy two conditions: they do not vary over time and they depend only on the properties of the current state s. These conditions indicate that every time system is in state s the transition probability to any other state will be always the same irrespective of how the state s was reached.

The transition probabilities also must satisfy the condition

$$\sum_{u \in \Omega} P(s \to u) = 1 \tag{4.41}$$

---

[1]The state space can be also considered as a mathematical set (Appendix B).

which states that given the system in state s the process must generate a new state. All of the above conditions do not imply that transition probability $P(s \to s)$ indicating that system stays in the same state s in the next move does not have to be zero. Interestingly if $P(s \to s) = 1$ then the Markov process is valid albeit trivial as the system stays in the state s. The Markov process defined in this section will generate a succession of states. When run for a long time with properly designed transition probabilities the states that are generated will appear in the chain with chances proportional to their probabilities and therefore will be the samples from the probability measure $p(s)$. Another requirement for Markov chain to reach a state in which samples can be obtained (reach an *equilibrium*) is requirement of *ergodicity*. The condition of ergodicity states that the Markov process should reach any state from any other state. The condition of ergodicity does not require that the state u should be reached from s in a single move but any number of moves can be used. In fact, for large systems with high number of states most of the transition probabilities between states will be zero. The reason for this is that it is much easier to handle Markov chains with sparse (most elements are zero) $P(s \to u)$.

4.4.2 DETAILED BALANCE

The final required condition for Markov process is the condition of *detailed balance*. This condition ensures that the chain reaches states with a chance that is proportional to the probabilities of these states. Often when that happens it is said that the Markov chain reaches the equilibrium. When the system is in equilibrium the rate of coming to a state s must be equal to the rate of going from the state s. This can be expressed by

$$\sum_{u \in \Omega} p(u)P(u \to s) = \sum_{u \in \Omega} p(s)P(s \to u) \tag{4.42}$$

where $p(s)$ and $p(u)$ are the probabilities of the chain being in states s and u which are equal to the value of the probability distribution defined on Ω. . Since $\sum_{u \in \Omega} P(s \to u) = 1$ by Equation (4.41) the Equation (4.42) becomes

$$p(s) = \sum_{u \in \Omega} p(u)P(u \to s). \tag{4.43}$$

The above guarantees that the process reaches equilibrium; unfortunately, it is an insufficient condition that states in the equilibrium are selected proportionally to $p(s)$.

To demonstrate this let's define *Markov matrix* $\mathbf{P}$ which is a *stochastic matrix* describing the stochastic evolution of the Markov process. By $w_s(t)$ we denote the probability that the Markov chain is in state s at Markov time t. The term *Markov time* or just time is used here to indicate the evolution of

Markov chain. It is a discrete variable that marks the position in the Markov chain of states. From the definition the Markov matrix is related to $w_s(t)$ as

$$w_s(t+1) = \sum_{u \in \Omega} P(u \to s)w_u(t) \qquad (4.44)$$

which can be compactly written in matrix notation as

$$\mathbf{w}(t+1) = \mathbf{P}\mathbf{w}(t). \qquad (4.45)$$

Using the introduced notation the equilibrium is reached when

$$\mathbf{w}(\infty) = \mathbf{P}\mathbf{w}(\infty) \qquad (4.46)$$

where $\mathbf{w}(\infty)$ indicates the equilibrium state as $t \to \infty$. Although Equation (4.43) indicates that the equilibrium is reached (expressed by Equation (4.46)) it does not rule out that the system is in *dynamic equilibrium* in which the $\mathbf{w}$ "rotates" around $\mathbf{w}(\infty)$. Such "rotation" is called *limit cycle* in which case

$$\mathbf{w}(\infty) = \mathbf{P}^n \mathbf{w}(\infty) \qquad (4.47)$$

where $n > 1$ is the length of the limit cycles. If Equation (4.43) is satisfied any number of limit cycles may exist. In such case the Markov chain may visit $\mathbf{w}$'s that are different than desired $\mathbf{w}(\infty)$ in which case the states visited (samples drawn) in dynamic equilibrium will not be proportional to the actual probabilities of those states.

To eliminate limit cycles, a stronger condition than the one expressed by Equation (4.43) can be enforced:

$$p(s)P(s \to u) = p(u)P(u \to s). \qquad (4.48)$$

The above is the condition of the *detailed balance* (DT). It can be quickly verified that the condition implies the equilibrium requirement stated by Equation (4.42) by applying $\sum_{s \in \Omega}$ on both sides of the equation. It also implies that limit cycles are impossible because if the system is in state s the rate of change to any other state is exactly the same as the rate of change back from this state. Therefore for every Markov move the probabilities are unchanged and dynamic equilibrium is impossible and for every Markov move $\mathbf{P}\mathbf{w}(\infty) = \mathbf{w}(\infty)$. The detailed balance implies a complete symmetry between states s and u. The Markov chains could have been constructed going from s to u and from u to s. This property is called *time reversibility*. The Markov chain that satisfies the detailed balance is called "reversible."

The transition probabilities of the chain that follow the detailed balance have to be of the form

$$\frac{P(s \to u)}{P(u \to s)} = \frac{p(u)}{p(s)}. \qquad (4.49)$$

If DT is satisfied and if the ergodicity condition is satisfied, the chain will reach an equilibrium and samples from $p(.)$ will be obtained.

4.4.3 DESIGN OF MARKOV CHAIN

We so far said that if we design an algorithm that switches from state to state
with probabilities of switching described by Equation (4.49) and that all states
are reachable from any other state it will produce samples from distributions
used in Equation (4.49). The actual implementation of this method may not
be obvious. There is no general recipe on how to create an efficient Markov
chain satisfying the detailed balance and ergodicity conditions. Every problem
has to be considered separately and a "trial and error" approach is used in
order to test various proposed algorithms. However, some general recipe for
the generation of the Markov chain can be constructed based on the fact that
the Markov move $s \to s$ is allowed by the condition of the detailed balance as
long as $P(s \to s) < 1$. This can be verified in Equation (4.49) as both sides of
the equation are 1. If the value of $P(s \to s)$ is equal to 1 the detailed balanced
is satisfied but the chain that will stay in s and therefore the ergodicity is
violated. The flexibility in selection of the probability $P(s \to s)$ provides the
following general scheme of constructing reversible Markov chains.

The design of reversible Markov chain algorithms can be divided into the
two following steps: (1) selection and (2) acceptance.

Selection In the selection step, given the state s some algorithm is used to
stochastically generate some new state u from Ω. In other words, if the chain
is in the state s the next state in the chain could be u subject to passing the
acceptance test in step 2. It is allowed to select the same state s as the chain
is currently in. The selection of the subsequent step is done stochastically
with *selection frequency* $S(s \to u)$. It indicates the relative frequency of
selecting the state u if the chain is repetitively in the state s.

Acceptance The acceptance step is performed after the selection step.
Therefore, the current state s is known and the selected state (proposed
state) u is also known. The *acceptance frequency* is the relative number of
times the u is accepted if the u was proposed while the chain is in state s.
This will be denoted as $A(s \to u)$.

Because these two stages are independent the transition probability is

$$P(s \to u) \propto S(s \to u)A(s \to u) \qquad (4.50)$$

This two-step algorithm produces either the same state s or a new state u
that is generated with the frequency described by $S(s \to u)A(s \to u)$.

The advantage of dividing the process of creation of the state in the Markov
chain of states is that it allows the freedom in choosing a convenient algorithm
for state generation. Using Equations (4.49) and (4.50) we have

$$\frac{S(s \to u)A(s \to u)}{S(u \to s)A(u \to s)} = \frac{p(u)}{p(s)} \qquad (4.51)$$

We identify in Equation (4.51) two terms $S(s \to u)/S(u \to s)$ and
$A(s \to u)/A(u \to s)$. For problems investigated by the Markov chain methods

essentially any algorithm can be used for state selection (subject to ergodicity condition) followed by such acceptance frequency ratio $A(s \to u)/A(u \to s)$ that ensures the chain obeys detailed balance.

Summarizing the approach, we have some algorithm that generates new states u given the current state s with frequency $S(s \to u)$ and then we accept those states with acceptance frequency $A(s \to u)$ which is chosen to satisfy Equation (4.51). When the algorithm attains the equilibrium, this will produce a chain of states which will appear to be drawn from distribution $p(.)$—a task that we are set to achieve. The subsequent states in the chain do not have to be different. A chain with very long chains of identical states is perfectly valid. Although this seems in disagreement with the property for reversible chains of generating states proportionally to their probability, the procedure can still generate states proportionally to its probability on average for infinite length chains. It follows that the states in the Markov chain are not independent and typically subsequent states are highly correlated. The aim of the design of a areversible Markov chain is to increase the acceptance probability as much as possible. This will The aim of the design of a reversible Markov chain is to increase the acceptance probability as much as possible. This will make the correlation between subsequent states smaller and allow generation of more independent samples per Markov move.

The increase in the acceptance frequency (ideally 100%) will typically not be possible within increasing complexity in the selection step and the algorithm has to be investigated for trade-offs between the complexity of the selection step and values of the acceptance frequency. Some of the most widely used algorithms are based on acceptance frequency equal to 100%. These algorithms include *Gibbs sampling* [32], *cluster algorithm* [107], and *Wolff algorithm* [111]. They do however require that selection frequency is equal to transition probability which may be difficult to find or computationally inefficient. In this book we exclusively use a two-stage algorithm based on Metropolis et al. [72] and later generalized by Hastings [41] described in the next section.

Example 4.4: Six-number Markov chain design

Suppose numbers $1 \ldots 6$ are considered. A distribution $p(.)$ is defined as the probability of a number that is proportional to that number (i.e., $p(1) \propto 1$, $p(2) \propto 2 \ldots$). This implies $p(1) = 1/21$, $p(2) = 2/21, \ldots$ after taking into account the normalization.

The goal is to design a Markov chain that generates samples from $p(.)$. As typically is the case, there are many possibilities of designing such reversible chain. We consider here only a few illustrative examples. The numbers are arranged around the circle as in Figure 4.7. First, we define the state space Ω which in this example has only six states corresponding to the six numbers. In the first example (left/upper diagram in Figure 4.7), the selection step is defined such that the chain moves only to two states corresponding to neighboring numbers. The are two possibilities of moving from any state. Assuming that both moves have the same frequency, possible moves have $S(s \to u) = 1/2$. Using Equation (4.51) the acceptance frequency ratio has to be equal to the ratio of probabilities of these states $p(.)$ in order for this chain to be reversible. An example ratio is $A(6 \to 1)/A(1 \to 6) = p(1)/p(6)$. We see that if we choose $A(s \to u) = p(u)$ it will satisfy the requirement and the chain using such acceptance frequencies will be reversible. This choice is by no means optimal but it will generate the reversible chains (see continuation of this example on Page 120 for optimal selection of acceptance ratio). It should be obvious that the chain is ergodic as all states are accessible from all other states. Using the description of the Markov chain provided the first few states were obtained as 6, 6, 6, 6, 6, 6, 6, 6, 6, 6, 6, 1, 1, 1, 1, 1, 6, 6, 6, 6, 6, 6, 6, 6, 6, 6, 6, 6, 6, 6, 6, 6, 6, 5, 5, 5, 5, 5, 5, 5, 5, 5, 5,...

The upper/right diagram in Figure 4.7 is similar to the one shown on the upper/left. The ergodicity is guaranteed. However, there is no direct route between 3 and 2 and there is a direct route between 2 and 3. The detailed balance condition cannot be attained between the state 2 and the state 3 and this chain is not reversible. The bottom left diagram in Figure 4.7 can also be reversible (subject to correct acceptance frequencies). The bottom right diagram is not reversible because state 2 is not reachable from any other state in the phase space.

4.4.4   METROPOLIS–HASTINGS SAMPLER

The Metropolis algorithm proposed in 1953 [72] was generalized by Hastings in 1970 [41]. The algorithm is derived here in the context of the general Markov chain design discussed in Section 4.4.3. By this design each Markov move consists of (1) selection and (2) acceptance steps. In the selection step, a state u is chosen based on selection frequency $S(s \to u)$ where s is the current state the chain is in. In the acceptance step, the selected state u is accepted with the frequency $A(s \to u)$. If not accepted, the chain stays in the state s. We will not introduce any changes to the selection step, but demonstrate how to design the acceptance frequencies $A(s \to u)$ to achieve the maximum efficiency algorithm. The good efficiency means that new states in the chain are frequently accepted.

Suppose we choose an algorithm for the state selection. The ratio of selection frequencies $S(s \to u)/S(u \to s)$ is known and for the chain to obey the detailed balance, the acceptance ratio must be

$$x \triangleq \frac{A(s \to u)}{A(u \to s)} = \frac{p(u)S(u \to s)}{p(s)S(s \to u)}. \tag{4.52}$$

The above condition is specified for the ratio of the frequencies and not on the actual values of A's. This gives some possibilities of selecting the acceptance frequencies so the actual values of A' are maximized. A straightforward choice

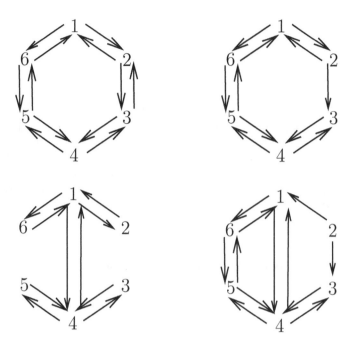

FIGURE 4.7 Four diagrams of possible selection schemes for creation of the Markov chain from Example 4.4.

for the frequencies $A(s \rightarrow u)$ and $A(u \rightarrow s)$ that satisfy Equation (4.52) is to request that

$$A(s \rightarrow u) = C\sqrt{x}$$
$$A(u \rightarrow s) = C\sqrt{x^{-1}} \tag{4.53}$$

where C is some constant. The constant should be such that the values of relative frequencies are in the range $[0, 1]$. At the same time C should be as high as possible to maximize the values of the acceptance frequencies. In order to find this value, we consider two cases:

Case $\sqrt{x} < \sqrt{x^{-1}}$ (or equivalently $x < 1$): The highest possible C is obtained from the second equation in Equation (4.53) by solving $C\sqrt{x^{-1}} = 1$.
Case $\sqrt{x} \geq \sqrt{x^{-1}}$ (or equivalently $x \geq 1$) : The highest possible C is obtained from the first equation in Equation (4.53) by solving $C\sqrt{x} = 1$.

   With that in mind, the acceptance frequency that satisfies both conditions is

$$A(s \to u) = \begin{cases} x & \text{if } x < 1 \\ 1 & \text{if } x \geq 1 \end{cases} \qquad (4.54)$$

where

$$x = \frac{p(u)S(u \to s)}{p(s)S(s \to u)}. \qquad (4.55)$$

The acceptance frequency of the reverse move $A(u \to s)$ is obtained from Equation (4.54) simply by switching labels s and u and the reader is encouraged to verify that $A(s \to u)$ and $A(u \to s)$ defined by Equation (4.54) satisfy the detailed balance condition. The Equation (4.54) defines the Metropolis algorithm. Together with the freedom of choosing the selection step in the creation of Markov chain this approach to creation of the Marrkov chains is known as the Metropolis–Hastings algorithm.

Example 4.5: Six-number Markov chain design (2)

This is the continuation of the example where numbers $1 \ldots 6$ are considered. A distribution $p(.)$ is defined as the probability of a number that is proportional to that number (i.e., $p(1) \propto 1$, $p(2) \propto 2 \ldots$). This implies $p(1) = 1/21$, $p(2) = 2/21, \ldots$ after taking into account the normalization. It is assumed that selection frequencies $S(s \to u)$ are all the same and equal to $1/2$.

Suppose we consider a selection scheme illustrated in Figure 4.7 in upper left diagram. There are twelve non-zero transitions $1 \to 2$, $2 \to 3$, $3 \to 4$, $4 \to 5$, $5 \to 6$, and $6 \to 1$ and corresponding reverse transitions.

Figure 4.8 presents a difference in efficiencies of the naˇve transition frequencies used in Example 4.4 with the Metropolis–Hastings acceptance frequencies derived in this section.

4.4.5 EQUILIBRIUM

The equilibrium is a stage at which the Markov chain produces samples from the underlying probability distribution $p(.)$ and therefore while in equilibrium the probability of finding in the chain the state s is proportional to the probability of this state $p(s)$. In an overwhelming majority of practical problems with a large number of dimensions there will be a very small subset of states which are probable and the Markov chain will sample these states. The much larger part of the phase space will consist of states that have probabilities that are vanishingly small which have almost non-existent chance to be sampled when the algorithm is in the equilibrium. Because the subspace of the probable states is small, it is unlikely that when the Markov chain is initiated (the first state is chosen) the state from probable states is selected.

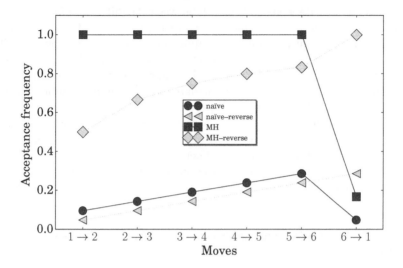

FIGURE 4.8 Acceptance frequencies for Markov moves defined in Figure 4.7 (upper/left diagram). The naïve frequencies are defined in Example 4.4. The Metropolis–Hastings frequencies are derived in Section 4.4.4. The "reverse" indicates the reverse moves to ones indicated on x–axis.

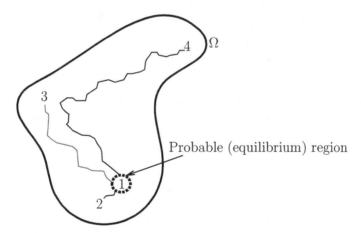

FIGURE 4.9 Burn-in times for different starting points 1, 2, 3, or 4 in the phase space.

Reaching the equilibrium

When the Markov chain is initialed with a low-probability state it will require some time (measured in Markov moves) to reach the probable states (see Figure 4.9). The time required to reach the equilibrium depends on the initialization and also depends on the path between the initialization state and the equilibrium region. Only a small region of the phase space Ω is visited by the Markov chain algorithm while in equilibrium. It is illustrated by a small dotted circle in Figure 4.9. The Markov chain can be initialed (first state is chosen) using any state from Ω. If initiated using a state from an equilibrium region (1 in Figure 4.9), the chain is immediately in the equilibrium but this is almost never the case for any practical problem. If initiated using some other states (2, 3, and 4 in Figure 4.9), the Markov chain requires some time to come to the equilibrium as illustrated in Figure 4.9. We will refer to this time as the *burn-in time* or *burn-in period*. There is no general prescription that can be used to estimate the length of the burn-in period and every problem has to be considered on a case-by-case basis. Consider the following example:

Example 4.6: MCMC and multivariate normal distribution – burn-in period

Suppose we would like to draw MCMC samples from a multivariate normal distribution $\mathcal{N}(\mathbf{x}|\boldsymbol{\mu}, \boldsymbol{\Sigma})$ (in Example 4.1 the samples were drawn directly). We assume that $\mathcal{N}(\mathbf{x}|\boldsymbol{\mu}, \boldsymbol{\Sigma})$ is the uncorrelated multivariate distribution and its variances are equal to 1. The matrix $\boldsymbol{\Sigma}$ is diagonal with values on the diagonal equal to 1. The number of dimensions is N and $\Omega = \mathbb{R}^N$. The vector of means $\boldsymbol{\mu}$ is equal to $[0, \ldots, 0]^T$. Suppose that $\mathbf{x}_0 = [20, \ldots, 20]$ is the first sample and the Markov chain is built starting from this sample.

In order to construct the Markov chain, the selection step has to be defined first. Two approaches to the selection of new states are investigated in this example. In approach 1, a sample from $\mathcal{N}(\mathbf{x}|\mathbf{0}, \sigma)$ is drawn directly and then added to the current state. The result is a new proposed state. If accepted in the acceptance step, this state will become the new state in the Markov chain. The $\mathcal{N}(\mathbf{x}|\mathbf{0}, \sigma)$ indicates the multivariate normal distribution with zero mean and diagonal covariance matrix with each diagonal element equal to σ. In the approach 2, a sample from the multidimensional uniform distribution is obtained and added to current state in the Markov chain. In each dimension, a sample from $\mathcal{U}(-\sqrt{3}\sigma, \sqrt{3}\sigma)$ was obtained. The term $\sqrt{3}$ was added to have the same 1-D variance of the selection step as in the Gaussian selection step.

Since the $\mathcal{N}(\mathbf{x}|\boldsymbol{\mu}, \boldsymbol{\Sigma})$ is "centered" around 0 for $\boldsymbol{\mu} = [0, \ldots, 0]$ it is expected that the equilibrium region will occur around a vector with all coordinates equal to zero. The starting state $\mathbf{x}_0 = [20, \ldots, 20]$ is far from the equilibrium region and the Markov chain requires several Markov moves in order to arrive to equilibrium region. Figure 4.10 shows that for some settings of the selection step the chain arrives very fast (e.g., $N = 1, \sigma = 1$) or very slow $N = 16, \sigma = 0.02$. More dimensions make the starting point further from the $\boldsymbol{\mu}$ and then the burn-in period is longer. How fast the system comes to equilibrium is also dependent on the size of the selection step. For this simple example the larger the selection step the faster the equilibrium is reached. However, this may not be the case for other sampling distributions.

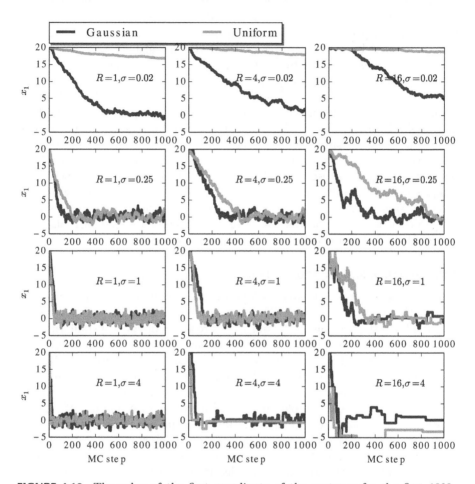

FIGURE 4.10 The value of the first coordinate of the vector **x** for the first 1000 steps of Markov chain sampling of multivariate normal distribution described in Example 4.6.

As demonstrated in Example 4.6, a poor choice of starting state and selection step can increase the burn-in time. One obvious suggestion is to select the starting state as close as possible to the equilibrium region. Sometimes, a good starting point will be a point estimate of the quantity of interest that can be quickly obtained. A poorly chosen selection step (or a poorly chosen starting state) leads to *poorly mixing* chains. A poorly mixing chain stays in small regions of the phase space Ω for long times (Figure 4.10 $R = 16, \sigma = 4$). On the other side a *well-mixing* chain seems to explore the phase space (Figure 4.10 $R = 1, \sigma = 1$).

One approach that should always be used for testing for the convergence to the equilibrium is to look at the *Markov trace* which is a plot of some QoI vs. the number of Markov step as illustrated in Figure 4.10. The trace shows the evidence of poor mixing and can also provide an estimate of the burn-in period. It must be cautioned however that the actual burn-in time can be much longer than suggested by the trace, but it is often obvious based on the trace that the burn-in is incomplete.

A more formal test for reaching the equilibrium is also available. For example, suppose we determined the point at which the equilibrium is reached and would like to test if the chain is at the equilibrium after this point. We could split the states after this point into two regions (e.g., first and last 20%) and then obtain average and standard deviation for some quantity derived from the samples and use methods of classical statistics to determine if the samples drawn from first and the last 20% of total number of samples come from the same distribution (e.g., the Z-score). For more details on this and other formal methods and discussion of these methods refer to Brooks and Gelman [14], Gelman and Rubin [30], and Robert and Casella [85].

Autocorrelation and MC standard error

When new states are generated in Markov chain they are likely to be correlated with the preceding state. This is because a Markov state is generated based only on the previous state and therefore a certain level of similarity between them is expected. Often, the Markov moves will only minimally change the state and therefore the subsequent states are almost identical. In the Metropolis–Hastings algorithm neighboring states can even be identical (perfectly correlated). As explained in the beginning of this chapter, ideally, independent Monte Carlo samples are needed for various estimations of means, integrals, sums, marginalizations, etc. Since the Markov chain generates correlated samples, a measure needs to be defined which would be indicative of the correlation between states. This measure would indicate the time (measured in number of Markov moves) needed for the system to arrive to a state that differs significantly and can be considered as an independent sample. We define this measure as the *correlation time* τ. One of the most straightforward use of τ is to perform *chain thinning*. Markov states are extracted every period that depends on τ. Extracted samples using chain thinning can be used as independent samples or samples with reduced correlation between them.

> **Important:** Although the Markov states are correlated, these correlations do not affect the means. Therefore, if the samples are obtained while the chain is in the equilibrium and the chain is sufficiently long, the estimates of the marginalized distributions, integrals, and sums as defined in Sections 4.1–4.3 are unbiased. That is, if an infinite number of samples are acquired, the MC estimates (averages over the samples) approach the true values that are being estimated. The correlation between Markov states

❚ will affect variances and covariances estimated from the MCMC samples.

It should be obvious that the small τ is a desirable feature of the Markov chain because it provides more independent samples from the same length chain. For the Markov chain the τ can be estimated by defining the *autocorrelation function* $\rho(t)$. The quantity t indicates the *Markov time* which in the case of *discrete Markov chains* is simply a number of states. For example $\rho(1)$ is the value of correlation function between a state and the next state, $\rho(2)$ is the value of correlation function between states separated by two Markov moves, etc. Clearly $\rho(0) = 1$ because any state is perfectly correlated (it is identical) with itself.

To define the autocorrelation function we assume that $\mathbf{x}_j$ are the Markov states enumerated $j = 1, \ldots, \infty$ and that $\phi_j = \phi(x_j)$ is some function of $\mathbf{x}$ where $\mathbf{x}$ is a vector quantity. The estimate of the autocorrelation function when R Markov states are available is defined as:

$$
\begin{aligned}
\hat{\rho}(t) &= \frac{\sum_{j=1}^{R-t} (\phi_j - \hat{\phi})(\phi_{j+t} - \hat{\phi})}{\sum_{j=1}^{R-t} (\phi_j - \hat{\phi})^2} \\
&= \frac{\sum_{j=1}^{R-t} (\phi_j - \hat{\phi})(\phi_{j+t} - \hat{\phi})}{\sum_{j=1}^{R-t} (\phi_j - \hat{\phi})^2}
\end{aligned}
\tag{4.56}
$$

For $R \to \infty$ the $\hat{\rho}(t) \to \rho(t)$. The value of the autocorrelation function falls between 0 and 1 where 0 indicates the lack of correlation (independence) and 1 indicates the complete correlation (all states are identical). From the definition (Equation (4.56)) the $\hat{\rho}(0) = 1$.

For Markov chains, the autocorrelation function is expected to fall off approximately exponentially and thus it can be modeled by

$$
\rho(t) \approx e^{-t/\tau}
\tag{4.57}
$$

where t is the time between samples and τ is the correlation time that we seek. Markov states obtained with time between them equal to $t = \tau$ are quite strongly correlated with the value of the correlation function equal to $1/e = 0.37$. It is reasonable to assume that t should be used such that it is at least two or more times τ. In practical applications we collect and save all Markov states (if possible) because the correlation time can only be estimated at the end of the MC simulation or at least until enough samples are available. Assuming the exponential model of the autocorrelation function, the estimation of the correlation time τ can be done by determining the values of autocorrelation function using Equation (4.56). A fit to those data points of the autocorrelation function provides the estimate of τ.

The other approach to compute τ from the autocorrelation function is to use

$$
\sum_{t=0}^{\infty} e^{-t/\tau} \approx \tau.
\tag{4.58}
$$

The approximation is better for higher values of τ. Using this approximation the estimate of τ can be obtained by summing the values of the correlation function as

$$\tau \approx \sum_{t=1}^{R} \hat{\rho}(t). \tag{4.59}$$

If the correlation time is estimated, it can be used to adjust computation of variances to account for the sample correlations. It was shown in Section 4.3 that variances of the MC-derived estimators are $\hat{var}(\hat{\Phi}) = \frac{1}{R-1}\sum_{j=1}^{R}(\phi_j - \hat{\Phi})^2$. Using this variance we defined the standard error as

$$\mathrm{se}^2(\hat{\Phi}) = \frac{1}{R}\frac{1}{R-1}\sum_{j=1}^{R}(\phi_j - \hat{\Phi})^2. \tag{4.60}$$

This expression assumes that samples ϕ_j are statistically independent which for the case of Markov chains will not be the case and Equation (4.60) has to be adjusted for this correlation.

It can be shown [74] that with correlated samples that standard error adjusted for correlation is

$$\mathrm{se}^2(\hat{\Phi}) = (1 + 2\tau_{eff})\frac{1}{R}\frac{1}{R-1}\sum_{j=1}^{R}(\phi_j - \hat{\Phi})^2 \tag{4.61}$$

where τ_{eff} is the *effective correlation time* after the chain thinning equal to $\tau/\Delta t$ where Δt is the interval that was used for chain thinning (samples with interval Δt samples were used in the calculations). Clearly Equation (4.61) becomes Equation (4.60) when $\tau = 0$.

4.4.6   RESAMPLING METHODS (BOOTSTRAP)

In previous sections the MC errors were discussed and it was explained how to compute standard errors of MC of quantities estimated using Monte Carlo methods. There are some cases where it is impossible or very difficult to estimate the error of some quantities using the methods described in Section 2.3.1 based on properties of the posterior. This can happen when the quantity of interest is derived in some complex way from samples and computing the posterior variance is either difficult or impossible[2]. Although not impossible, the calculation of errors of those estimates is not straightforward using the direct methods because the correlations between estimates of the distributions. The

---

[2]In Section 6.2.6 we calculate the error of the Bayes risk used to quantify imaging systems.

bootstrapping and *jackknifing* are two types of methods that are straightforward and that can save a great deal of effort in calculation of errors compared to direct method.

The bootstrap method is essentially a *resampling method*. To show the utility of this approach, we define an estimate of some quantity $\hat{\Phi}$ that depends on R samples ϕ_j of this quantity. Ideally the samples should be independent but in practice it transpires that the bootstrap method will often provide a reasonable result for correlated samples.

The bootstrap method is based on generation of B new sets of samples (*bootstrap sets*). One of those new B bootstrap sets is obtained by drawing R samples with repetitions from the original R samples. This implies that about 63% of samples in the new bootstrap sets will be duplicates. The estimate of the quantity $\hat{\Phi}$ is calculated from the bootstrap sets the same way as for the original estimate. The number of B should be as large as computational resources allow in order to minimize any possible effect of random sampling when creating the bootstrap sets. The bootstrap estimate of the standard error [25] is

$$\mathrm{se}^2(\hat{\Phi}) = \frac{1}{B}\sum_{b=1}^{B}\hat{\Phi}_b^2 - \left(\frac{1}{B}\sum_{b=1}^{B}\hat{\Phi}_b\right)^2 \qquad (4.62)$$

where $\hat{\Phi}_b$ is the estimate of quantity Φ from samples in bootstrap set b.

5 Basics of nuclear imaging

In nuclear imaging (NI) in medicine, the electromagnetic radiation is used to obtain information about various physiological functions. The radiation originates inside the patients and ideally the detectors of NI cameras detect as many as possible of the photons originated form within the body. The energy of electromagnetic radiation used for imaging is large enough to penetrate the tissues. However, if the photon energy is high (for example 2 MeV per photon) and almost no *attenuation* occurs in the body, the detection of such high-energy radiation is difficult (the matter becomes "transparent") and hardware needed for detecting high-energy radiation impractical in a clinical setting. As a result of this trade-off, radiation energy is used in the range of 50-1000 keV in nuclear imaging. The energy is sufficient for photons to leave the body and at the same time to be detected using relatively inexpensive and small-form detector technology.

The radiation is delivered to subject bodies by the administration of a small amount of *tracer* that is labeled with *radionuclides*. A tracer labeled with a radionuclide is called a *radiopharmaceutical*. The radiation is created when radiopharmaceutical decays and produces either directly or indirectly photons that are later detected by the NI camera. In certain radioactive decay, photons (at least one) are the direct product of the nuclear reaction that occurred. In an indirect photon production, the radioactive decay produces a positron (positively charged electron) which annihilates producing a couple of secondary photons. Because the principle of detection of the direct and indirect photons is different, two imaging modalities are identified. The single photon emission modality that includes planar and tomographic (single photon emission computed tomography or SPECT) scans is concerned with imaging using direct photons and Positron Emission Tomography (PET) concerned with imaging positrons (or indirect photons). In Sections 5.3.3.1, 5.3.3.2 and 5.3.4.1 we provide details that differentiate those nuclear imaging modalities.

The medical applications of nuclear imaging have a variety of diagnostic uses in the clinics. In addition to clinical applications, the nuclear imaging modalities (SPECT and PET) are the backbone of the field of *molecular imaging* that is used to better understand molecular pathways and is extensively utilized in pre-clinical trials with animals to test new drugs and therapies. We provide examples of the medical and other applications in Sections 5.5.1 and 5.5.2 and for thorough review of uses of clinical and pre-clinical uses of nuclear imaging the reader should refer to Phelps [79], Rudin [86], and Weissleder et al. [109].

5.1 NUCLEAR RADIATION

5.1.1 BASICS OF NUCLEAR PHYSICS

The atomic and nuclear physics is concerned with the description of processes that involve the smallest building blocks of biological matter—the atoms. These blocks interact between each other by many complex and not intuitive processes and certain simplifications need to be used in order to describe these processes in limited space in this book. For a more thorough coverage of nuclear physics processes the reader is encouraged to consult Griffiths [36] and Krane [60].

5.1.1.1 Atoms and chemical reactions

The basic building blocks of all matter are *atoms*. An atom comprises of a dense and heavy nucleus that carries a large majority of the total weight and a *cloud of electrons* that surrounds the nucleus. The nature of electrons exhibits properties of both particles (that intuitively can be illustrated by billiard balls) and waves which are mathematical functions describing probabilistically the locations and momenta of electrons. The probabilistic description is governed by the rules of quantum mechanics. The atom (nucleus and electrons) is held together by electromagnetic forces between a positively charged nucleus and negatively charged electrons. The size (radius) of the nucleus is in the order of 10 fm (1 fm = 10^{-15} m), and the size of the atom is in the order 10^5 fm and therefore the nucleus is tiny compared to the extent of the electron cloud. An analogy of the pea in the center of the football stadium can be used to illustrate the relative sizes of a nucleus and an atom (electron cloud).

Although the nucleus is tiny compared to the volume occupied by the electron cloud, the mass and energy of an atom is concentrated in the nucleus. The nucleus is built from positively charged *protons* and electrostatic neutral *neutrons*. The mass of protons and neutrons is approximately equal and is about 1840 times higher than the mass of electrons. Therefore, the gross mass of the atom is concentrated in the nucleus. An analogy can be used of a goose down pillow to illustrate the electron cloud in which case the nucleus would be of the size of the tip of a needle in the middle of the pillow weighting 1 tonne (1 tonne = 1000 kg).

Chemical reactions involve primarily the outermost electrons from the electron clouds in chemically reacting atoms. The interaction of electron clouds of different atoms in chemical reactions does not significantly affect the states of their nuclei and therefore nuclear reactions are not affected by the chemical composition of atoms in chemical molecule (chemical molecule is a group of atoms bounded together by electrostatic forces and sharing electron clouds).

The type of chemical reaction the atom can undergo is dependent on the number and configuration of electrons in the electron cloud which in turn is dependent on the number of protons (positively charged *nucleons*) because the number of electrons is equal to the number of protons and therefore depends indirectly on the nucleus type. Although the chemical properties are not directly dependent on the nucleus by a convention, different atoms are categorized in terms of the chemical properties by the number of protons in the nucleus.

The number of protons is the *atomic number* of the atom and is indicative of the chemical properties of atoms. Atoms with the same atomic number are known as *chemical elements* to emphasize their almost identical chemical properties. The number of neutrons in the nucleus does not significantly affect the chemical properties of the atom. It does, however, affect the properties of the nuclei and will directly affect the nuclear reactions. The number of protons (atomic number) added to the number of neutrons in the nucleus is the *mass number* of the atom. The same *element* (short for chemical element) can have many different numbers of neutrons in the nucleus. These variants of the atoms with the same atomic number but different mass number (different number of neutrons) are the *isotopes*.

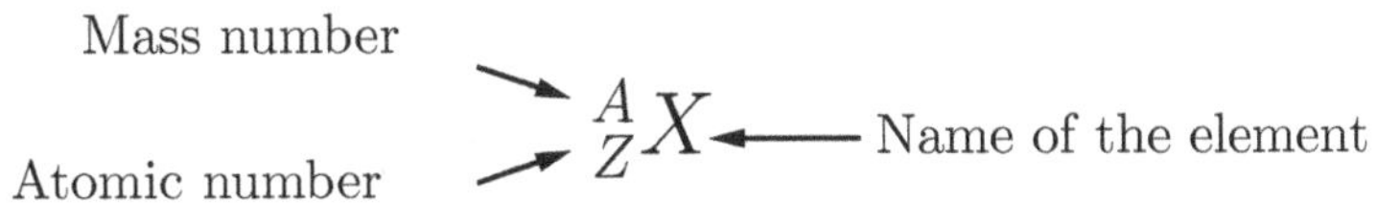

$$\text{Mass number} \searrow \quad {}^{A}_{Z}X \longleftarrow \text{Name of the element}$$
$$\text{Atomic number} \nearrow$$

FIGURE 5.1 Atomic notation.

In atomic notation (Figure 5.1), chemical elements are indicated by the atomic number Z and the name of the element X. This is redundant information as just one of Z or X unambiguously determines the chemical element. Sometimes the notation will be used in which the atomic number is omitted $({}^{A}X)$ due to this redundancy. The mass number A identifies the isotope of element X.

5.1.1.2 Nucleus and nuclear reactions

Since the nuclei are built of neutrons and protons that are very close together, and considering that protons are positively charged, they exert a large repelling electrostatic force, and if no other interaction exists nuclei that consist of two or more protons would fall apart due to those repelling electrostatic forces. The attractive force behind the fact that nuclei hold together is called *residual strong force* or *nuclear strong interaction*. The strong interaction force has a very short range and therefore is significant only when *nucleons*, which is a name for particles that comprise the nucleus, are every close together. The nucleons have to be as close as the diameter of either a proton or a neutron for the strong force to "kick in." The strong nuclear force is created between nucleons by the exchange of particles called *mesons*. If nucleons are close enough for the strong interaction to occur, the exchange of mesons starts, and the particles will stick to each other overpowering the repelling interactions (electrostatic). However, they have to be close enough otherwise the strong force is too weak to make them stick together, and repelling forces would make the particles move apart.

The adjective "residual" in the name of the strong force (residual strong force) comes from the fact that when nucleons are consider together (when they form the nucleus) the sum of their masses is smaller than the sum of masses of nucleons considered separately. The difference (the residual) can be considered as the energy that binds the nucleus together. Figuratively speaking, the *binding energy* has to be delivered to the nucleus in order to take the nucleons apart.

In order for protons and neutrons to coexist in the nucleus, an equilibrium between repulsive and attractive forces must be in place. The presence of neutrons in the nucleus helps to reduce the repulsion between protons because they have no charge they do not add to the repulsion already present between protons. At the same time neutrons are sources of strong forces for the nucleus since they participate in the meson exchange. These meson interactions create enough strong force to overcome proton mutual repulsion and force the nucleons to stay bound together.

There are many possibilities that certain numbers of protons and neutrons can be arranged in the nucleus. Some of the arrangements will be impossible, some will be very unstable, and some will stay together for a very long time. The number of protons and neutrons as well as the arrangement of them in the nucleus define the *nuclide*. For a certain nucleus with specified number of protons and neutrons, the energies corresponding to different arrangements of protons and neutrons in the nucleus are different. The arrangement with the smallest energy is the *ground state* of the nucleus (nuclide corresponding to the ground state of the nucleus). All other states with higher energies are *excited* or *metastable* nuclides. If a nucleus is excited it very quickly transforms (in the order of 10^{-12} seconds) to the ground state which is accompanied by emission by the nucleus of a *gamma* "prompt" photon. For metastable states the transformation to the ground state is slower and takes about 10^{-9} seconds or more to transform to the ground state. The distinction between excited and metastable states is somewhat artificial, as there is no conceptual difference between these types of nuclear excitations except by the average time that is needed to transform to the ground state.

From the definition, the *stable nucleus* does not spontaneously undergo *radioactive decay* or at most it cannot be observed in long observation time that it does. For each known chemical element, there usually is at least one stable isotope; however, there are elements that do not have stable isotopes. There are a total of about 250 configurations of the number of proton and neutrons that combined form stable nuclides. Figure 5.2 presents the *decay scheme* for unstable isotope of oxygen. The transition from unstable ^{15}O to stable ^{15}N (Figure 5.2) is accompanied by releasing 2.75 MeV of energy that

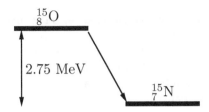

FIGURE 5.2 Decay scheme diagram for ^{15}O.

in the case of ^{15}O is in the form of particles "ejected" from the nucleus.

If a nucleus is not stable (*radionuclide*), it changes its configuration (number of neutrons and protons) and such process is termed as *radioactive decay*. A *radioactive nucleus* can decay to a different nucleus which can be stable or not. The nucleus that is the result of the decay is called the *daughter product*. If the daughter of the decay is unstable it can decay to another radionucleus or a stable nucleus, etc. The decay of nuclei is a probabilistic process in which the decays can occur at any time with the same chance equal to *decay constant* λ. Because of that the time at which the nucleus will decay is governed by the exponential probability law. We have used this fact in the derivation of the counting statistics in Chapter 3. For applications in nuclear imaging, λ can be assumed constant and independent on time and conditions the nucleus is in (pressure, temperature, etc.). This assumption can be questioned [4]; however, the possible deviations are very small (if any) and mostly occur for very short or very long lived radionuclides and therefore are irrelevant for nuclear imaging.

5.1.1.3 Types of nuclear decay

There are several types of nuclear decays which result in the emission of *gamma rays* and/or some particles such as *alpha* or *beta*. In nuclear imaging we are mostly interested in the types that produce photons in the energy range of 50 keV to 1000 keV. Photons from this energy range can be detected by the camera and based on those detections the original distributions of the locations of the radionuclide can be inferred. We briefly describe the types of nuclear decay relevant to the nuclear imaging. We only discuss γ emissions which are photons emitted as a result by the nucleus changing its energy state. The photon emissions by orbital electrons (*x-rays*) changing their energy states (quantum states) are not thoroughly discussed in this short introduction. The topic is covered in more details in many textbooks (e.g., Cherry et al. [18]).

β^- **decay** Illustratively speaking the β^- radioactive process is a transformation of one of the neutrons in the nucleus to the proton, electron, and antineutrino. The *antineutrino* and *neutrino* are elementary subatomic par-

ticles that do not have any electric change. They likely have a mass but
the upper bound on the value of this mass is so tiny that it has not been
experimentally validated yet. The antineutrino and neutrino interact with
other matter only through *weak* [1] and *gravitational* forces and are extremely
difficult to detect and study experimentally. Because of this reason we will
not be concerned with neutrinos in imaging.

When one of neutrons in the nucleus is converted to a proton, the atomic
number of the nucleus is increased by one (as is the positive charge) and
the mass number remains the same. It is represented in standard nuclear
notation as

$$\ _{Z}^{A}X \xrightarrow{\beta^-} \ _{Z+1}^{A}X' \tag{5.1}$$

The electron produced by β^- reaction does not have a high significance
in nuclear imaging; however, after the β^- reaction some nuclei are left in
an excited state and emit prompt gamma photons that can be used for
imaging. An example of such nuclei is $_{54}^{133}$Xe with the decay diagram shown
in Figure 5.3.

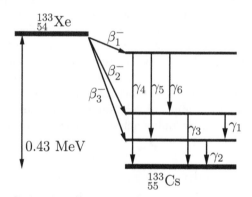

FIGURE 5.3 Decay scheme diagram for ^{133}Xe which is a β^- emitter. There are
several prompt gammas (γ_1 through γ_6) that are produced.

Isomeric transition and internal conversion After the decay occurs, the
daughter nucleus is usually in an excited state. Some of those exited states
may be very long lived (even years). Such "long-lived" excited nuclides are
in the *metastable* state. The decay from matastable state is called *isomeric
transition* (IT). The occurrence of IT is accompanied by the emission by the
nucleus of γ photons or by the process of *internal conversion* (IC). In the
IC process the metastable (and excited) nucleus interacts with an electron

---

[1]Weak nuclear force is one of the fundamental interactions found in nature. Radioactive
decays are governed by weak interactions.

cloud of orbital electrons and transfers the excess of energy to one of the electrons. Because of this, the IC electron is ejected from the atom carrying out the kinetic energy (less the electron binding energy) it gained in the IC process. The IC does not result in emission of γ photons. The ^{99m}Tc nuclide that undergoes IT and IC is the most frequently used nuclide in nuclear imaging.

Electron capture The last two types of nuclear reactions that have relevance in nuclear imaging decrease by one the number of protons in the nucleus. Therefore, the atomic number is reduced by one as well. In the *electron capture* (EC) the electron cloud interacts with the nucleus and one of the electrons is "absorbed" by one of the protons. Figuratively speaking one of the electrons is captured by the nucleus. The results of the EC are: (1) the vacancy in one of the orbital electrons, (2) the neutrino is emitted, (3) and the nucleus is in an excited or metastable state. Therefore, the gamma rays may also be emitted as a result of the EC.

β^+ **decay** The β^+ decay results in the production of a *positron* (antiparticle of electron) in the nucleus which is ejected with some kinetic energy from the atom. The neutrino is also produced and emitted by the nucleus. After the emission of the positron and neutrino, the nucleus can be left in an excited or metastable state and therefore some prompt gammas can also be emitted. In the nuclear imaging the most interesting is the emitted positron which after losing its kinetic energy due to collisions with electrons combines with one of the electrons and *annihilates*. In the annihilation reaction, the masses of the particle and antiparticle and their kinetic energies are fully converted to massless energy of photons that are created. For electron-positron annihilation in nuclear imaging the annihilation occurs when the electron and positron are almost at rest. The energy carried by two photons in the laboratory frame is only slightly higher than the combined mass of electron and positron that accounts for the residual kinetic energy of electron and positron during annihilation. In the first approximation, however, it can be assumed that two photons that are produced have equal energies and propagate in exactly opposite directions due to the energy and momentum conservation. The energy of each photon is 511 keV so the sum (1022 keV) is equal to the total rest mass of positron and electron. Technically 511-keV annihilation photons are not γ rays because they are created outside the nucleus. The PET imaging is based on detection of those photons.

Figure 5.4 shows decay diagram of one of the most frequently used nuclide for PET imaging ^{18}F (for more details on applications of ^{18}F, see Section 5.4). The ^{18}F decays to stable nuclide ^{15}N through either electron capture or β^+ decay. The maximum kinetic energy that the positron receives is equal to 633 keV which is the difference between the mass of parent nuclei of ^{15}O and masses of ^{15}N, an electron and a positron (1022 keV). On the first look it is puzzling that 1022 keV is used for creating a single positron that is ejected from the nucleus. One would expect only 511 keV to be used for creating the positron. The reason for this is that, in the nuclear reaction, positron and electron must be created. The positron is ejected from the nucleus and the electron is "assimilated" by the nucleus. Therefore, the process requires creation of the mass equal to mass of two beta particles.

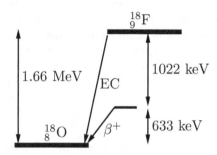

FIGURE 5.4 Decay scheme diagram for ^{18}F.

5.1.2   INTERACTION OF RADIATION WITH MATTER

From the point of view of imaging, the interaction of photons created in nuclear processes with the matter is important because of two reasons. The goal of imaging is to determine the distributions or locations of the sources of the radiation. In order to do so, the photons created at the source have to reach the detectors (detectors are discussed in Section 5.2). The interaction of the radiation with the object that is imaged will *attenuate* some of the photons and not all of the original photons reach the detector. Some photons may be deflected in their path (*scattered*) and when they reach the detector they may be coming from a different direction than the direction of the source. The scatter can occur in the object and in any structure that surrounds the nuclear imaging camera. It is not impossible (albeit insignificant) to think about a photon emitted by the object scattered by the building across the street and coming back and be detected.

The interaction of photons with the matter is also important because photons can only by detected indirectly through the interaction of photons with detectors. This creates a trade-off mentioned already in the beginning of the chapter. On one hand the attenuation of photons should be reduced as much as possible, but on the other hand we would like the detectors to register as many of the incoming photons as possible. These are two conflicting goals. Since the attenuation and scatter are functions of photon energy and materials that they interact with, the optimal range of photon energies to be used for human nuclear imaging is around 50 keV to 1 MeV. For lower energies in this range, less expensive and difficult to build detectors can be used but scatter

and attenuation are substantial. For higher energy photons, the situation is
the opposite.

We provide a short introduction to physics of photon interactions with
the matter important for understanding processes of photon attenuation and
detection in nuclear imaging.

5.1.2.1   Inelastic scattering

The photons may be deflected in their paths through the matter if they interact with electron clouds through so-called *Compton scatter* interaction.
This interaction is also called inelastic scattering. In Compton scattering (Figure 5.5) the electron gains some of the energy from the incident photon and
if the gained energy is higher than the binding energy it is knocked out from
the atom (*recoiled electron*). The outer shell electron can be considered as
a free electron which makes the mathematical modeling of this interaction
much simpler. The scattered photon is deflected in its path by an angle ψ
with respect to its original direction and loses some of its energy. The recoiled electron gains some speed $\vec{v}$ (kinetic energy). The photon energy after
scattering is equal to E_{sc}.

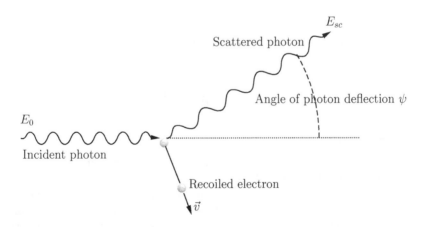

FIGURE 5.5   Diagram of Compton scattering. Incident photon with energy E_0 is
scattered on an outer shell electron.

From the conservation laws of energy and momentum the relation between
scattering angle ψ and the energy of scattered photons can be derived and is
expressed by:

$$E_{sc} = \frac{E_0}{1 + \frac{E_0}{511 \text{ keV}}(1 - \cos\psi)} \tag{5.2}$$

It is clear from this equation that for $\psi = 0$ the energy of scattered photon is unchanged compared to E_0, so it indicates no deflection and no interaction. Also the minimum energy of scattered photon is for $\psi = \pi$ radians which corresponds to so-called *backscatter*. The incident photon is "bounced back" as a result of the collision with the electron. The energy of the backscatter photon is equal to $\frac{E_0 511\text{keV}}{511\text{keV}+2E_0}$. For the case of annihilation photons with $E_0 = 511\text{keV}$ the backscatter photon energy is equal to $511 \text{ keV}/3 = 170 \text{ keV}$. The Equation (5.2) applies to the interaction of a photon with a free electron and should be slightly modified when modeling interaction with electrons bounded in atoms. Due to electron binding the *Doppler broadening effect* occurs in which the Compton-scattered energy spectrum is widened due to pre-Compton-interaction momentum of the electron [110]. The strength of the Doppler broadening effect depends on the energy of the photon and is stronger for low energies of the gamma photons. The angular distribution (differential cross-section) $d\sigma/d\psi$ of scattered photons on free electron is described by the Klein-Nishina formula [58], which is a function of the photon energy E_0 and scattering angle ψ.

$$\frac{d\sigma}{d\psi} = const \times \left\{ P(E_0, \psi)^3 + P(E_0, \psi) - P(E_0, \psi)^2 \sin^2\psi \right\} \sin\psi$$

where

$$P(E_0, \psi) = \frac{1}{1 + (E_0/511\text{keV})(1 - \cos\psi)}.$$

5.1.2.2 Photoelectric effect

In the *photoelectric effect*, all the incident photon energy is absorbed by an electron bounded in the atom (photon disappears) and an orbital electron is ejected. The energy of incident photon must be greater than the bounding energy of the *photoelectric electron* for the photoelectric interaction to occur. The photoelectric electron is ejected from the atom with the kinetic energy equal to the difference between the energy of incident photon and binding energy of the electron.

5.1.2.3 Photon attenuation

The *photon attenuation* is one of the most important effects in nuclear imaging and must be taken into account for analysis of the NI data. In this book we put an emphasis on the statistics, and we do not discuss approaches for accounting for photon attenuation; however, the attenuation must be taken into account in any real application.

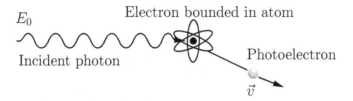

FIGURE 5.6 Diagram of the photoelectric interaction. The energy of the incident photon is transferred to one of the electrons. The electron is ejected from the atom with the kinetic energy equal to the difference between E_0 and the binding energy of the electron.

The original gamma ray is considered to be attenuated when it interacts with the matter through any possible interaction. The Compton scatter (Section 5.1.2.1) and photoelectric effect (Section 5.1.2.2) are the two main interactions that photons undergo while passing through matter in the range of energies used in nuclear imaging. In addition, the photons can also undergo *Thompson scattering*. In Thompson (also known as classical or elastic) scattering the photon is deflected without losing any energy. The momentum is conserved as the change in photon direction is compensated by change in the momentum of the atom. The probability of occurrence of Thompson scatter is low compared to other two interactions and plays a marginal role in photon attenuation and nuclear imaging.

Although the nature of attenuation is discrete, as it occurs between photons and atoms which both are discrete quantities, the attenuation can be parametrized by a continuous quantity because of an enormous number of atoms in matter makes discrete description impossible. We advocate in this book the discrete treatment of radioactive nuclides because their number is many orders of magnitude smaller than the number of atoms in subjects, but to represent the attenuation medium the continuous approximation is more reasonable. We define linear attenuation coefficient κ in units of cm^{-1} that describes an infinitesimal chance δ that a photon is attenuated when it travels through infinitesimal thickness of some attenuator. In other words, the greater the κ, the stronger the attenuator. The linear attenuation coefficient is the function of the photon energy and types of atoms that the matter is composed of as well as the thermodynamic state of the matter. We will be concerned with imaging under normal atmospheric pressure and room temperature and use photon-attenuation properties for such conditions.

The linear attenuation coefficient is a convenient quantity to mathematically describe photon attenuation. Suppose we consider attenuation between points 1 and 2, on Figure 5.7. The shaded object represents the attenuator. The distance between these points can be divided into very thin layers dr and

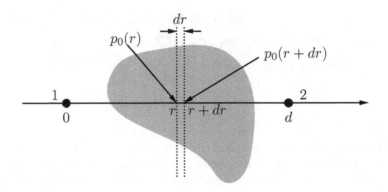

FIGURE 5.7 Photon attenuation.

by the definition of the linear attenuation coefficient we have that

$$\delta(r) = \kappa(r)dr \tag{5.3}$$

where $\delta(r)$ is a chance that a photon emitted from 1 is attenuated in the infinitesimal layer at r and κ is the attenuation coefficient at r. Suppose we consider a chance that the photon is not attenuated and reaches location r $p_0(r)$ (Figure 5.7). If this happens, the chance that it is not attenuated at the location $r + dr$ is $p_0(r + dr) = p_0(r)(1 - \delta(r)) = p_0(r)(1 - \kappa(r)dr)$ (using Equation (5.3)). We rearrange the terms and have

$$\frac{p_0(r+dr) - p_0(r)}{p_0(r)} = -\kappa(r)dr. \tag{5.4}$$

For $dr \to 0$

$$(\log p_0(r))' \, dr = -\kappa(r)dr. \tag{5.5}$$

Finally, integrating both sides with integration limits $r = 0$ and $r = d$ and taking into account the initial condition $p_0(0) = 1$ we have

$$p_0(d) = \exp\left(-\int_0^d \kappa(r)dr\right) \tag{5.6}$$

The concept of photon attenuation is illustrated in Example 5.1.

Example 5.1: Photon attenuation

We consider a point source (all photons are emitted from a mathematical point) of photons and three detectors positioned far from the point source. We assume perfect detectors, which means that if a photon hits the active area of the detector ΔS it is registered with certainty. Each detector has the same active area ΔS which is small compared to the squared distance to the point source. Suppose the distances of the detectors from the point source are r_1, r_2, and r_3. Then, the chance ϵ that a photon is detected (hits the detector) is:

$$\epsilon = \frac{\Delta S}{4\pi} \left(\frac{1}{r_1^2} + \frac{1}{r_2^2} + \frac{1}{r_3^2} \right) \tag{5.7}$$

This is illustrated in Figure 5.8(A).

In Figure 5.8(B) the source of photons is placed in the center of attenuating sphere with a constant linear attenuation coefficient κ and radius d. Therefore each photon before it is detected has to pass through d thickness of attenuating medium and therefore the chance that a photon is detected is

$$\epsilon = \frac{\Delta S}{4\pi} \left(\frac{1}{r_1^2} + \frac{1}{r_2^2} + \frac{1}{r_3^2} \right) \exp\left(-\kappa d\right) \tag{5.8}$$

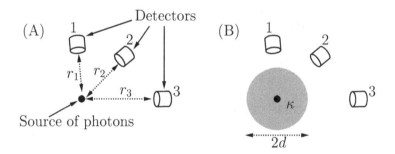

FIGURE 5.8 (A) A point source of photons is located at r_1, r_2 and r_3 distance from three small perfect detectors. (B) The point source is placed in the center of attenuating sphere with radius d.

5.2   RADIATION DETECTION IN NUCLEAR IMAGING

This section is a brief summary of a technology used for the detection of photons in the range of 50 to 1000 keV used in nuclear imaging. The discrete counts corresponding to single photons are detected as opposed to some aggregate signal created by many photons. Such detectors are referred to as *photon counting detectors*. The term *count* will frequently be used that indicates a detection of one photon or detection of more photons which were the result of a single decay. For example of β^+ decay simultaneous detection of both annihilation photons constitutes the count (more details of imaging of β^+ emitters are given in Section 5.3.4.1).

The radiation detectors should have a high *detector efficiency*. The detector efficiency describes the ability to stop the incoming photons. A higher detector efficiency indicates a higher chance that a photon interacts with the detector. It is therefore desired to have the detectors build from materials that are dense and have a high attenuation for gamma rays. The active area of detectors should also be as large as possible to cover the large solid angle around the object. Achieving a high efficiency is associated with a high cost (more expensive materials in larger quantities) and therefore trade-offs between the price and efficiency are made when selecting detector materials and scanner geometry for nuclear imaging.

Another important characteristic of detectors is how fast the detector is ready for another count after the detection. This characteristic sets limits on the maximum *count rate* achievable with the camera. The count rate is the number of detected counts per second. When detecting the high-energy photon, it generates secondary lower-energy photons (Section 5.2.2) or electron-hole pairs (Section 5.2.1). The number of those secondary photons or electron-hole pairs is indicative of the energy of incident photons and therefore the statistical error associated with this number translates to the uncertainty in the measurement of the energy of the original high-energy photon.

The large majority of nuclear imaging scanners are based on *scintillation* materials which are used to create the secondary photons and are discussed in Section 5.2.2. Semiconductor detectors are introduced in Section 5.2.1.

5.2.1   SEMICONDUCTOR DETECTORS

Semiconductors are poor electrical conductors. However, when a gamma ray hits the semiconductor material, it generates ionizations. If an electric field is applied, electric current is induced, and the generated charge is collected (Figure 5.9). The collected charge is proportional to energy of incident gamma ray and therefore the semiconductor detectors can be used to energy-selective counting. In nuclear imaging this is an important factor as photons that are Compton scattered can be discriminated from unscattered photons based on the energy.

The most popular semiconductor material used in nuclear imaging is Cadmium Zinc Telluride (CZT). This semiconductor can be operated at the room temperature which is a major advantage over other semiconductors such as Silicon or Germanium. The atomic number of Cadmium (48), Zinc (30), and Telluride (52) is also relatively high making it an excellent material to stop photons with energies that are used in nuclear imaging. In Table 5.1 the density of CZT can be compared to other commonly used materials (scintillators) used for nuclear imaging. The schematics of the operation of CZT detector are shown in Figure 5.9.

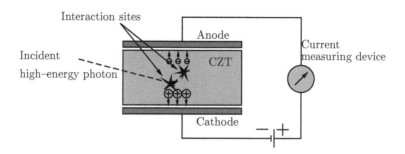

FIGURE 5.9 The CZT is placed in an electric field. Incident photons create electric charge (electron-hole pairs) in the CZT material.

5.2.2 SCINTILLATION DETECTORS

When a high-energy photon interacts with atoms of scintillator, it indirectly produces a few hundred to a few thousand UV (ultra-violet) visible-light photons. The number of those photons is proportional to the energy deposited by the high-energy photon and therefore scintillation detectors can be used for the measurement of the energy of the incident high-energy photons and energy-selective counting. For a detailed physical mechanism involved in this process refer to Knoll [59].

Radiation detectors made from scintillators are called *scintillation detectors*. The most common scintillation crystal used in nuclear imaging is sodium iodine doped with tellurium (NaI(Tl)). The production of this crystal is relatively inexpensive. It also has good imaging properties which makes it an excellent material for a scintillator used in gamma cameras (Section 5.3.3.1). The NaI crystal is not well suited for the detection of annihilation photons (511 keV) because of a relatively low density. For the detection of 511 keV photons, the Bismuth germinate $Bi_4Ge_3O_{12}$ (BGO) or Cerium-doped Lutetium Yttrium Orthosilicate (LYSO) is frequently used. Various properties of the scintillators and semiconductors are listed and compared in Table 5.1.

The general operation of the scintillation detectors is presented in Figure 5.10. When scintillator is hit by a high-energy gamma ray, it produces many (hundreds) low-energy photons. These photons are converted to an electrical signal and amplified. The final electrical signal is measured and is indicative of the original energy of the high-energy photon (Figure 5.10). The UV and visible light produced by scintillator is converted to electronic signal using traditional photomultiplier tubes (PMTs), avalanche photodiode (APD), and silicon photomultipliers (SiPM) discussed in the next section.

TABLE 5.1

Properties of scintillator and semiconductor materials used in nuclear imaging

Property	NaI(Tl)	BGO	LYSO	CZT
Density (g/cm^3)	3.67	7.13	7.40	5.85
Effective atomic number[a]	50	74	66	49
Photon yield[b] (per keV)	38	8	20-30	N/A
Hygroscopic	yes	no	no	no

[a] Effective atomic number is the weighted average of the atomic number of the components of the material.
[b] Photon yield is the average number of low-energy photons produced by the gamma ray per keV.

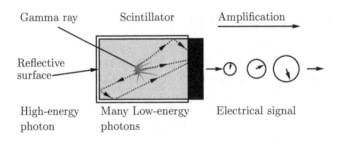

FIGURE 5.10 Scintillation radiation detector.

5.2.2.1 Photomultiplier tubes

The basics of the operation of the photomultiplier tube (PMT) is illustrated in Figure 5.11. The basic idea is to convert photons created by the gamma ray in the scintillators to measurable electrical signal. This is done by using a *photocathode* made from a material which hit by low-energy (UV or visible) photons ejects electrons through the photoelectric effect. Typically few low-energy photons are needed to generate one electron, but the ratio depends on the photocathode material and energy of incoming visible light/UV photons.

In the PMT, the electrons ejected from the photocathode are accelerated in the electric field maintained between the photocathode and the first *dynode*. Accelerated electrons striking the first dynode produce secondary electrons which in turn are accelerated in the electric field maintained between the first dynode and the second dynode. The voltage difference between electrodes in the PMT is about 100 Volts. The process of acceleration and multiplication of electrons is repeated as many times as the number of dynodes. There are in the order of 10 dynodes in PMT. The last electrode in the PMT is called the *anode*. Elections reaching the anode produce easily detectable electric pulse which is associated with gamma photon striking the scintillator. The schematic of the operation of the PMT is shown in Figure 5.11.

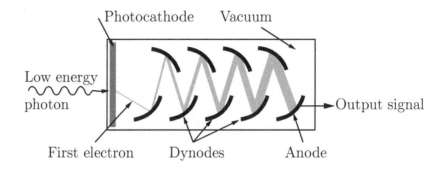

FIGURE 5.11 Diagram showing basic principles of the PMT. The gray line illustrates the stream of electrons.

5.2.2.2 Solid-state photomultipliers

The PMTs are a well-established technology and have been used as the photomultiplier of choice in nuclear imaging. The high PMT gain is achieved at the expense of high voltage in the order to 1000 Volts which requires quite costly power supplies and large form factor. The PMTs are highly sensitive to magnetic fields which limits the use of this technology in some applications in imaging.

These adverse characteristics of PMT lead to development of silicon solid-state multipliers. The basic idea of operation of solid-state multipliers is similar to semi-conductor detectors discussed in Section 5.2.1. Here however the incoming photons have low energies and the material commonly used to build the solid-state multipliers is the silicon. The basic example of such device is the PIN diode which can used for photodetection. The PIN diode does not provide any signal multiplication and when the low voltage is applied to the PIN diode, it conducts electricity only when it is exposed to incident photons generating electron-hole pairs.

The avalanche photodiode (APD) is an example of solid-state amplifiers. All light-sensitive silicon diodes operate under the reversed polarity (*reversed bias*). A string electric field is used in APDs so the initial ionizations (electron-hole) created by incident low-energy photons are multiplied by accelerated electrons. The multiplication factor of the number of ionizations is around 100 for about 200 Volts applied. Higher voltages can also be applied resulting in higher amplifications. While the gains are in general much smaller than those obtained by PMTs for which it can be in the order of 10^6, the sensitivity for creation of the electron-pair is much higher than sensitivity of creating electrons at the PMT photocathode. The drawback of the APDs is high-noise in the output signal which translates into a poor energy resolution. There is also an important trade-off of noise of the resulting signal and timing. The thicker the APD, the lower the noise in the amplified signal. On the other hand, the time needed for created charge (electrons and holes) to reach the electrodes increases resulting in a worse timing characteristics of the APD (worse achievable count rates).

A different mechanism is used with another type of the solid-state PM, the silicon photomultipliers (SiPM). When a high-intensity electric field is generated in the silicon beyond *breakdown point* (approx. 5×10^5 V/cm) the charge created by incident photons is accelerated to a point that it triggers self-perpetuating ionization cascade in the silicon volume that is subjected to the high electric filed. This Geiger cascade generates a measurable current which is independent of the initial number of photons or their energy which triggered the cascade. Such a process is called "Geiger discharge" and is similar to the process used in Geiger–Müller detector. The cascade is quenched by using a resistor attached to the photodiode in series. The flow of the current will decrease the voltage below the breakdown point and the Geiger discharge stops. The reduction in current resets the photodiode to the initial state (see Figure 5.12). The photodiode in the Geiger mode provides only a binary output and no information can be drawn about the initial number of photons that triggered the cascade.

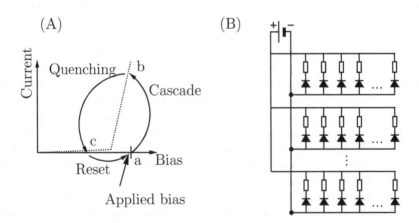

FIGURE 5.12 (A) Basic operation of the Geiger-mode photodiode. (a) The initial state. Incident photon rapidly raises the current (b) which triggers drop in bias due to use of quenching resistor and the photodiode resets to the initial state. (B) Schematic representation of the array of photodiodes with quenching resistors (microcell).

Since every photodiode creates a signal that is independent of the number of photons that triggered this signal, the proportionality to the initial photon flux is achieved by creating a dense arrays of optically independent photodiodes each with its own quenching resistor. This couple is referred to as *microcell.* The number of microcells can be in the order of thousands per mm^2. The microcells are connected such that signal is summed (Figure 5.11(B)) and the total output is considered the output from the SiPM device. The sum of the binary signals obtained from each microcell is therefore capable to provide information about the total optical photon flux that reached the SiPM.

Table 5.2 presents comparison of various photomultipliers.

TABLE 5.2
Properties of various photomultipliers [90]

	PMT	PIN	APD	SiPM
Gain^a	10^6	1	10^2	N/A
Voltage	High	**Low**	High	**Low**
Temperature sensitivity	**Low**	**Low**	High	High
Form factor	Bulky	**Compact**	**Compact**	**Compact**
Magnetic field sensitivity	Yes	**No**	**No**	**No**
Timing^b	Slow	Medium	**Fast**	**Fast**

^{a} Gain is defined as the charge at the output divided by the charge created by the incident photons.
^{b} Timing defines the ability of photomultiplier to reset the PM to the initial state ready for conversion and amplification of another photon flux.
Note: Desired characteristics are in the bold font.

5.3   NUCLEAR IMAGING

In nuclear imaging (NI) the spatial distribution of the pharmaceutical nuclei inside the body of the patient or some other object or volume is analyzed. This analysis is based on photons emitted in the subject, but observed (detected) outside the imaged object. These photons are either gamma radiation or annihilation photons. This spatial distribution of radiotracer can be monitored over time providing insights into physiological state of the organisms *in vivo.* We will refer to such imaging as the *functional imaging* as the function of the organism is assessed. The NI techniques are inherently functional, as they monitor how tracers are metabolized or used and do not focus on the structure of the body. The functional imaging can only be performed on living organisms and NI would not provide any information for dead subjects. Other imaging modalities such as x-ray Computed Tomography (CT), Magnetic Resonance

Imaging (MRI) or Ultrasound (US) are not inherently functional. The main function of these modalities is investigation of the structure; however, these methods have been extended also to investigation of functions. These imaging modalities are not covered in this book and interested readers are directed to Bushberg et al. [16], Dendy and Heaton [23], and Huda [44].

In order to introduce basic principles of nuclear imaging, we first define some new concepts. The most general description of the subject of imaging is described by locations of all radioactive nuclei that are in the subject assuming that they are attached to the tracer. This is the most general description of the state of nature which is not directly observable. In medical applications the number of such nuclides is astronomical and could reach 10^{12}.[2] It should be obvious that obtaining any information about such vast number of unknowns is impossible. Therefore, we simplify the description of the object by subdividing the volume into non-overlapping equal in size *volume elements*. The simplest and most flexible and easy-to-use volume element is a *voxel* which we will utilize in this book. For simplicity of the exposition of the theory, we assume in this book that all voxels in the imaged volume have the same size, but variable size [97] and even overlapped volume elements [70] can also be used. The voxel representation is in fact quite general. For example, it can be used as an approximation of continuous space when the voxel size approaches zero. We will not be interested in the actual exact locations of nuclides inside the voxel but only in the number of nuclides inside voxels. This concept is illustrated in Figure 5.13. The most general description of the object is specified by locations of radioactive nuclei Figure 5.13 (left). This is approximated by volume-element representation Figure 5.13 (right) where only the numbers of nuclei inside the volume elements (here inside the squares (pixels)) are considered. In Figure 5.13 (left), the number of nuclei is converted to a color using a color scale. Color scale is a lookup table that for a number of nuclei per voxel assigns a color.

In nuclear imaging we will be concerned with getting an inference about the number of radioactive nuclei (which are typically attached to tracers) or the number of decays of those nuclei per voxel. Let's put it more formally:

> **Definition:** The spatial distribution of radioactive nuclei is defined by the number of radioactive nuclei per voxel. This quantity will be referred to as the *radiodensity*. Symbol r will be used to indicate this value. In Chapter 3 we already have used symbol r to indicate the number of radioactive nuclei in voxels.

Many authors use the term *radioactivity* or simply *activity* to quantitatively describe the rate of number of decays per second. Although this value is not necessary for description of the counting statistics as it was not used in

---

[2]This is the total number of radioactive nuclides, not the number of decays.

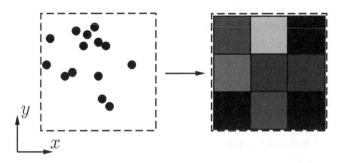

FIGURE 5.13 Voxel representation of imaged volumes.

Chapter 3, we provide the definition because it is a key quantity used in the imaging field. How this quantity relates to the theory used in this book is important. We will use symbol a to denote the activity. The radiodensity and activity are closely related. For example, if r is the number of nuclei in a voxel, the activity of this voxel is

$$a = \lambda r \qquad (5.9)$$

where λ is the decay constant. The value of activity a is proportional to discrete quantity r (Equation (5.9)) and it is frequently assumed that it is a continuous quantity for large r. However, it should be clear that for a sample of radioactive material the activity can take only discrete values. The activity is typically measured in units of *Becquerel* (Bq). Unit of radioactivity Bq corresponds to 1 decay per second. Another somewhat obsolete but still frequently used unit of activity is 1 *curie* (Ci) which corresponds to the radioactivity of 1 gram of *radium* isotope ^{226}Ra and

$$1 \text{ Ci} = 1000 \text{ mCi} = 1000000 \text{ } \mu\text{Ci} = 3.7 \times 10^{10} \text{ Bq.}$$

In nuclear imaging mCi and μCi units are used because the amounts typically used in imaging patients are in the order of mCi and μCi.

Example 5.2: Radiodensity vs. activity concentration

Suppose we consider a 1 voxel that contains 10^9 nuclei of ^{82}Rb. The radiodensity is 10^9 . This quantity does not indicate the frequency of radioactive disintegration but it is indicative of the concentrations (how densely the radioactive nuclei are packed in voxel) of the nuclei that can potentially undergo nuclear decay. To obtain activity, it is sufficient to multiply the radiodensity by the *decay constant* for given nuclide (in this case rubbidium 82). The decay constant is a property of the nucleus. In Chapter 3 the decay constant is discussed in details. The decay constant for ^{82}Rb nuclide is 0.00919 s^{-1} and therefore the activity is

9.19×10^6 Bq or 0.25 mCi. Table 5.3 presents the relation between radiodensity and radioactivity for several commonly used nuclides in nuclear imaging.

TABLE 5.3

Relationships between activity and radiodensity for common nuclides used in nuclear imaging

Nuclide	a [mCi]	a [MBq]	λ [minutes^{-1}]	Half-life [minutes]	r [1]
^{18}F	1	37	6.31×10^{-3}	110	5.86×10^9
^{15}O	1	37	3.40×10^{-1}	2.04	1.09×10^8
^{11}C	1	37	3.41×10^{-2}	20.3	1.09×10^9
^{82}Rb	1	37	5.51×10^{-1}	1.26	6.71×10^7
^{13}N	1	37	6.96×10^{-2}	9.97	5.32×10^8
^{99m}Tc	1	37	1.92×10^{-3}	360	1.93×10^{10}
^{123}I	1	37	8.74×10^{-4}	793	4.23×10^{10}
^{201}Tl	1	37	1.58×10^{-4}	4.38×10^3	2.34×10^{11}
^{22}Na^a	1	37	8.44×10^{-9}	1.37×10^6	4.38×10^{15}

^{a} The ^{22}Na is not used to image living organisms because of a very small value of decay constant and a high radiodensity is needed for imaging. A high radiodensity will result in a high radiation dose delivered to the patient. The nuclide is used however in quality control and calibration of nuclear imaging scanners [87].

Note: Source of nuclide data: National Nuclear Data Center, Brookhaven National Laboratory (http://www.nndc.bnl.gov/)

5.3.1   PHOTON-LIMITED DATA

The radioactive nuclei in the object can decay producing gamma rays and these gamma rays can be detected by the imaging camera. We define a *radioactive event* or simply *event* as the process of radioactive decay that produces directly or indirectly (e.g., through annihilation) one of more gamma rays that can interact with detectors of the imaging camera. If such interaction occurs, *a count* is registered. Thus, when a count is detected it corresponds to an event that occurred in imaged volume. However, when event occurs it may or may not produce a count in the imaging camera. We always refer to "event" in the context of the imaged volume and to "count" in the context of detection.

The distinguishing characteristic of nuclear imaging is the acquisition of photon-limited data (PLD) which is a list of counts. The raw data is acquired on the count-by-count basis. We assumed in our approach that the number of counts is limited (up to a 10^9) and can be handled (acquired and saved) by current computers count by count. In some other imaging modalities that use radiation as x-ray CT that would be impossible because of the much larger number of counts.

We will refer to data acquired and stored in this format as the *list-mode* data. Ideally the list-mode data constitutes raw outputs from the detectors (e.g., signals from PM tubes). These requirements of data storage in such formats are high and therefore seldom used. Frequently, a decision about the initial interaction location inside the detector (e.g., in which crystal does the interaction occur) is made by the Anger logic or other hardware and therefore only the result of the decision crystal number or location within the crystal is stored in the list-mode file. Other information about detected events such as detection times or output signal strengths (indicative of energy) are also saved in the list-mode data.

In the past the photon-limited data were acquired in so-called *projections* which are the histograms of the detected event. If counts were detected with similar characteristics, they were summed in a histogram bin corresponding to those characteristics and only the number of counts in each bin was considered. With an advent of modern imaging technologies this mode of data acquisition became mostly obsolete and typically the data are acquired and stored in the list-mode format and only reduced to the histogram form if needed. Figure 5.14 presents the schematics of the data acquisition in photon-limited imaging. The event timeline illustrates radioactive events occurring in the imaged volume (Figure 5.14 (upper line)). Some of those events are detected and produce counts shown on the count timeline (Figure 5.14 (middle line)). For each of the counts the region of response (see Section 5.3.2) is constructed which corresponds to possible locations of event origins. In general, each count has a different ROR but with a similar intricacy (Section 5.3.2).

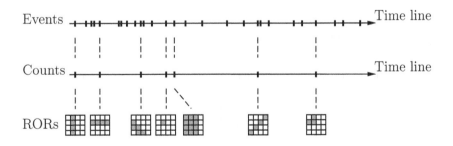

FIGURE 5.14 Event and count timeline.

5.3.2   REGION OF RESPONSE (ROR)

Upon detection of a count by the camera, a *region of response* (ROR) has to be defined (Figure 5.14). The ROR is a region (collection of voxels) in the imaged volume where the event could have occurred. Here we provide two extreme examples of such regions. Suppose that we operate an imaging camera which based on the detected count, determines the exact x, y, and z coordinates of the radioactive event that generate the count. Of course this is a quite unrealistic scenario but illustrates the perfect event localization characteristics of the imaging camera. Since xyz are known, the voxel that contains these coordinates is also known. The opposite scenario is when a detection of a count occurs and nothing can be said about the location of origin of the event that generated the count. For such a case any voxel can contain the origin of such an event with equal chance (assuming equal voxel sizes). Obviously, imaging cameras with such characteristics would be quite useless for obtaining any information about the spatial distribution of radioactive nuclei.

The RORs obtained by realistic cameras are somewhere between these two extremes. Since this concept of ability of localization of the origins for detected counts is one of the most important characteristics of the imaging camera we summarize it with the term *ROR intricacy* and denote it as ζ. We provide some intuitive arguments which should illustrate the concept. We assume that if an imaging system (each ROR) provides perfect localization, the intricacy of the imaging system is 0 (the best). For more realistic scenarios we measure the intricacy of the ROR by the information entropy defined in bits. So, for example, the worthless imaging system that for each detected count assigns equal change to each voxel as the origin of this count the intricacy $\zeta = \log_2 I$.

To put it more formally, the intricacy of a ROR corresponding to detector element k is defined

$$\zeta_k = -\sum_{i=1}^{I} \frac{\alpha_{ki}}{\sum_{i'=1}^{I} \alpha_{ki'}} \log_2 \frac{\alpha_{ki}}{\sum_{i'=1}^{I} \alpha_{ki'}}. \tag{5.10}$$

and the intricacy of the imaging system $\bar{\zeta}$ is simply the average intricacy defined as

$$\bar{\zeta} = \frac{1}{K} \sum_{k=1}^{K} \zeta_k \tag{5.11}$$

Here we use an average over all the detector elements. Other definitions of posterior system intricacy are discussed in Chapter 6.

Using these definitions the intricacy of $\bar{\zeta} = \log_2 I$ corresponds to a complete lack of knowledge about where detected counts could have originated (useless imaging system). The imaging system with a lower intricacy should be preferable over systems with a higher intricacy. The concept is illustrated with the following example:

Example 5.3: Intricacy

Suppose that volume is divided into I voxels of equal size as shown in Figure 5.15. Suppose further that counts are detected and the possible origin locations for the radioactive event for each count can be localized in a voxel, on a line, or a plane through the volume in any of three perpendicular directions. A voxel, a line and a plane in voxel representations are shown in Figure 5.15(A), (B), and (C). The voxel representations of a point, line, or plane in this example constitute the RORs. Figure 5.15(D) shows the entire volume as the ROR which corresponds to worse imaging system.

The reconstruction area is defined as a volume built from I voxels. Therefore the size (the edge) of the cube is $\sqrt[3]{I}$ voxels. For simple 1D ROR shown in 5.15(B) there are $\sqrt[3]{I}$ voxels with value of system matrix for detector element k equal to some value a and $I - \sqrt[3]{I}$ voxels that 0. It follows that $\sum_{i'=1}^{I} \alpha_{ki'} = a\sqrt[3]{I}$. As before we define $0 \log_2 0 \equiv 0$. Therefore

$$-\sum_{i=1}^{I} \frac{\alpha_{ki}}{\sum_{i'=1}^{I} \alpha_{ki'}} \log_2 \frac{\alpha_{ki}}{\sum_{i'=1}^{I} \alpha_{ki'}} = -\frac{a\sqrt[3]{I}}{a\sqrt[3]{I}} \log_2 \frac{a}{a\sqrt[3]{I}} = \frac{1}{3} \log_2 I \qquad (5.12)$$

Using similar approximate arguments the intricacy for ROR shown in 5.15(C) and (D) are $\frac{2}{3} \log_2 I$ and $\log_2 I$.

The ROR intricacy scales linearly with dimensionality of the ROR.

5.3.3 IMAGING WITH GAMMA CAMERA

In this introduction we discuss three implementations of the nuclear imaging devices and we start with the most popular device used in medical applications which is the imaging with the *gamma camera*.

5.3.3.1 Gamma camera

The basics of the design of the gamma camera has not changed since it has been introduced in the 1950s [2, 3]. Sometimes the gamma camera is referred to as "Anger camera" due to its inventor Hal Anger. Over the years it became the most widely used nuclear imaging device in hospitals around the world.

The gamma camera is based on the concept of using a large scintillation crystal and many PMTs to convert the light generated in the crystal to electric signals which are indicative of the location of interaction of the gamma ray with this large slab of crystal as well as the energy of the gamma photon. In scintillation gamma cameras used today the sodium iodine dopped with thallium crystal is used almost exclusively. This is because of the ease of

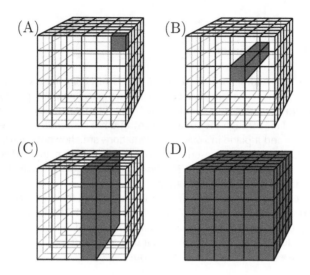

FIGURE 5.15 The voxelized imaging volume (A). An example of ROR (darkened voxels) (B) that corresponds to line. (C) and (D) show RORs that correspond to a plane and a volume.

growth of the large volumes of the crystal as well as a relatively low cost of manufacturing.

Figure 5.16 presents the heart of any gamma camera system. Although not shown in the figure, the NaI(Tl) crystal is encased because it is hygroscopic. When gamma ray hits the crystal it generates visible light that propagates to photocathodes of all the PMTs. In general, each PMT produces different strength signals but the PMTs that are closest to the location of the interaction will have the strongest signal. Ideally, these signals should be saved to the raw list-mode file for further analysis. Typically, however, the signals are analyzed and the most likely location of the interaction of the gamma photon with the crystal is estimated at time of data acquisition using the *Anger logic* [3]. Also the energy is estimated based on the sum of signals from all PMT and both, the location and energy, is saved in the list-mode file.

The estimation of the location of the interaction within the crystal during the acquisition is suboptimal because it is based only on measurements (PMT signals) and therefore prone to statistical errors. The actual location of the interaction is correlated with data from other counts and based on this correlation the estimation can be improved.

In the Anger cameras used in the hospitals, the NaI(Tl) crystals used are either round or rectangular with diameter of 50 cm and to 60 × 50 cm, respectively. The thickness of the crystal used varies from 6 to 12 mm. There is a trade-off as the thicker crystal provides higher sensitivity (more gamma photons are stopped) but at the same time the localization of the actual interaction of the photon with crystal is more uncertain. The number of PMTs used varies from 25 to 100 PMTs.

The schematics of a gamma camera is presented in Figure 5.16, which shows the integral part of gamma camera NaI(tl) crystal slab and PMTs attached to it. The source of gamma photons is on the left of the camera as shown in Figure 5.16. The NaI(tl) crystal is surrounded from all sides except the side with PMTs by a reflective material (mirror) that reflects the light generated in the NaI(Tl) crystal by the gamma photon. The crystal slab and PMTs are separated by a light-conducting material (light guide) as presented in Figure 5.16.

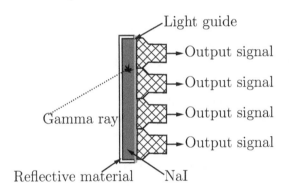

FIGURE 5.16 Gamma camera.

Obviously the gamma camera as shown in Figure 5.16 would be quite useless as its intricacy would be very high. That is, upon detecting a count the location of the event (the origin) would be highly uncertain because we would not know what direction the photon that generated the count came from.

In order to improve count-localization properties of the gamma camera, the *collimation* is used. The camera is shielded from the sides and the back and the *collimator* is positioned in the front of the camera (Figure 5.17). The goal of the collimator is to restrict directions of photons that are detected and therefore increase the count-localization properties. This improvement in localization comes at the cost of sensitivity as the use of collimator stops more than 99% of gamma photons that would otherwise be detected as counts. The basic task of a collimator is illustrated in Figure 5.17.

The collimator and shielding is built from some high-Z value material as lead or tungsten that has a high value of attenuation coefficients for high-energy photons. In Figure 5.17 the collimator is depicted as a series of thick lines (plates in 3D) in front of the gamma camera. The photons with paths

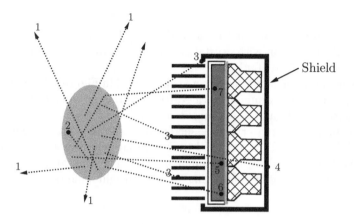

FIGURE 5.17 Gamma camera with a collimator and the object (shaded oval). Only gammas numbered 5, 6, and 7 produce a count on the detector.

that are not approximately parallel to the collimator plates are likely to hit it and be absorbed (photons with number 3 in Figure 5.17). The gamma camera is positioned at some angle vs. the object and the large majority of photons are emitted in directions that do not intercept the camera (photons identified by number #1 in Figure 5.17). Some of the photons will be absorbed in the object itself (#2) and some will be absorbed in the collimator or the shielding (#3 and #4 in Figure 5.17). The count occurs when a photon path is approximately parallel to the plates of the collimator (#5) in the example shown in Figure 5.17. The count will also occur when a photon penetrates the collimator and interacts with the crystal. Sometimes the direction of the photons is changed by the Compton or Thompson interactions and the photon is counted by the detector (#7) because the original direction was changed. We provide descriptions of only a few possibilities of the hypothetical gamma photons.

There is a variety of collimator designs that can be used with gamma camera (Figure 5.18). The most widely used is the *parallel hole* collimator. For the parallel-hole collimator photons that are coming with directions perpendicular to the surface of this collimator have the highest chance to be detected. Other collimator designs include pin-hole, converging, and diverging collimator. Ideally, all those collimators provide ROR with intricacy of approximately $0.33 \log_2 I$ because the localization of the origin of the event is reduced to a line.

Modeling the ROR by a line for the parallel-hole collimators (and other collimators as well) is a unrealistic simplification. Because of the finite size of the holes in the parallel-hole collimator, photons are counted that are coming with angles slightly smaller than 90° (see photon #7 in Figure 5.17). In addition, if for a count a possibility that the count corresponds to a Compton scatter is taken into account, the origin of the events can actually be located anywhere in the volume. This will further increase the intricacy of the ROR.

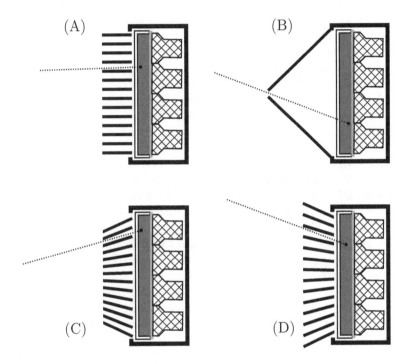

FIGURE 5.18 Four most common types of collimator used in nuclear imaging. There are parallel-hole (A), pin-hole (B), converging (C), and diverging (D) collimators. Dotted line shows the path of a detected photon.

5.3.3.2 SPECT

In the previous section, the gamma camera was introduced. If the gamma camera is placed facing the object, the information obtained is not sufficient for inferences about the 3D distribution of nuclides in the object because from the single view provides only a so-called *projection*. Images of objects at different depths are overlapped in the projection. For some diagnostic tasks (decisions) the information about the 3D volume distribution of the nuclides may be critical and needed. The gamma camera has to be moved and acquire counts from different views to obtain *tomographic* information about the 3D object. Whether the data is sufficient to obtain insights about the 3D distri-bution is dependent on the number of acquired counts, the intricacies of the

RORs corresponding to those counts, the relative configuration of the RORs, and finally it is also dependent on the imaging task that we would like to achieve. The interplay of those factors is difficult to characterize in general.

We refer to single-photon emission computed tomography (SPECT) as the imaging technique in which the gamma camera is moved around the patient and acquires counts at different projections. Traditionally, the gamma camera is moved to different locations and acquires projections. In modern cameras this term is obsolete as the data is acquired in the list-mode format. The most common mode of data acquisition is to rotate the camera around the patient. This rotation can either be accomplished by step-and-shoot or continuous mode. Counts are saved to the list-mode stream along with other indicators acquired during the scan such as time stamps, camera-gantry rotation, gamma camera position and tilt, etc. The modern SPECT scanners are equipped with two gamma cameras that can be positioned with a relative angle to each other and speed up data acquisition (Figure 5.19). Each of the two heads is an independent gamma camera. With this acquisition setup, the system needs only a 90° rotation to acquire counts from all directions in the 2D plane perpendicular to the axis of rotation. Dotted lines (Figure 5.19) illustrate hypothetical paths of gamma photons which can be absorbed by the object (attenuated) (#2) or scattered and then detected (#7). These are two examples of many possible interactions.

In SPECT, it is assumed that distribution of the radioactive nucleotides inside the object does not significantly change during acquisitions. This assumption may be violated because the distribution of the tracer is governed by underlying physiological processes which by nature are dynamic. Although there are techniques developed that account for these changes (see the review in Gullberg et al. [37]) they have not been popularized. Therefore, the SPECT imaging is mainly used for imaging of static tracers which distribution does not change or the change is not significant.

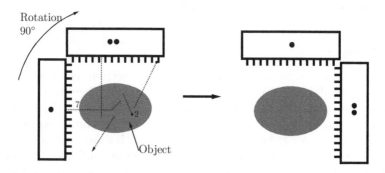

FIGURE 5.19 Data acquisition in SPECT using two-head scanner.

In recent years there are new SPECT cameras on the market which are stationary and no rotation is required. The imaging done by these cameras is focused on a specific organ (heart or brain) in pre-clinical imaging [7] or clinical imaging [15, 28, 37, 98].

5.3.4 POSITRON EMISSION TOMOGRAPHY (PET)

The goal of PET imaging is to determine distribution of radionuclides that emit positrons which subsequently annihilate producing 511 keV photons. Photons are detected by a large number of detector elements that surround the patient. These detector elements form typically a ring around the scanned object. The diameters of PET cameras vary depending on the application and for clinical scanners they are about 80 cm.

Positron (e^+) is ejected from nucleus with a kinetic energy that is largely lost due to interaction of the positron with other atoms (mainly electrons of those atoms) before it annihilates with an election (e^-) (Figure 5.20). The distance traveled by the positron after decay and before annihilation is the positron range indicated in Figure 5.20. The *positron range* varies and can be as large as 1.65 cm for ^{82}Rb (the maximum range) [83]. The annihilation produces typically two 511 keV photons traveling in opposite directions. The angle between photon directions in laboratory frame is always slightly different than 180° (*non-collinearity of the annihilation photons*) because the momentum of the two 511 keV photons has to be equal to sum of momenta of the election and positron which are never zero at annihilation. The positron range and non-collinearity contribute to uncertainty in localization of detected events.

5.3.4.1 PET nuclear imaging scanner

During acquisition of PET data, the object is positioned inside the ring of PET detectors which is schematically illustrated in Figure 5.22. Currently, the most popular approach to PET scanner design is to use the block detectors as the basic elements of the scanner. This block detector is shown in Figure 5.21. The high-energy (gamma) photon interacts with the scintillation material of the block and produces a number of low-energy photons (visible light) which are converted to electronic signal and multiplied by a set of photomultipliers. In Figure 5.21 four such multipliers are shown but other configurations with a higher number of PMs can also be used. Based on relative signals from the PMs it is determined using Anger logic (relative strength of the signals coming from different PMTs) where within the volume of the crystal block the actual interaction occurred. To increase the accuracy of the identification of the location of the interaction, the block is cut into *crystals* which have about 4–5 mm in size. In Figure 5.21 the block of scintillation material is cut into 36 crystals.

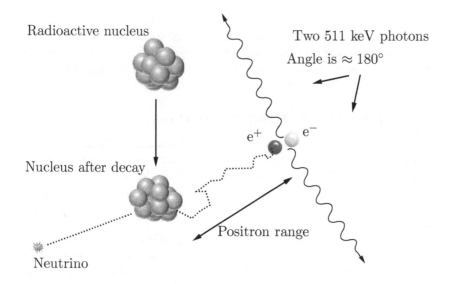

FIGURE 5.20 Schematics of β^+ nuclear decay with indirect production of 511 keV photons.

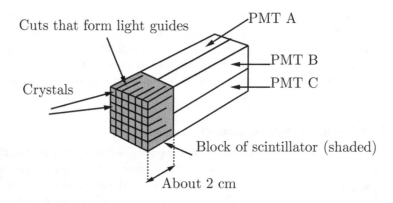

FIGURE 5.21 Block detector used commonly in PET cameras.

The number of blocks used in the ring of a PET scanner is in the order of 70 to a 100 (there are 24 blocks visualized in Figure 5.22). Each block is cut into about 150 to 200 crystals. Therefore, there are roughly about ten thousand crystals in the ring of PET detectors. Modern PET scanners are built from a series of up to five of such rings which amount to a total of about fifty thousand crystals. Considering that a annihilation event produces two annihilation photons that can be detected in those crystals, the number of ways an annihilation event can be detected by such a scanner is in the order of 10^9. This is a much higher number than the total number of counts detected by the scanner which is in the order of 10^8.

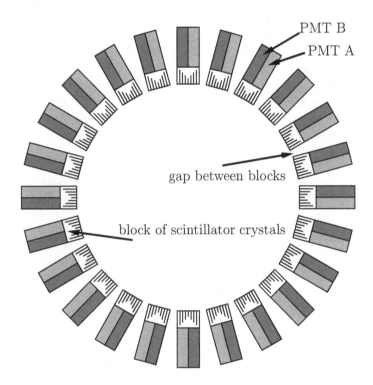

FIGURE 5.22 PET camera built from 24 PET block detectors.

5.3.4.2 Coincidence detection

Two photons produced by the annihilation may interact with the crystals of the PET camera. This is shown in Figure 5.23(A). If photons arrived during the *coincidence window*, they are detected *in coincidence* and, ideally, photons are assumed to be the result of the same annihilation. The situations where both annihilation photons are detected by the camera are quite rare because the angular coverage of the object by the PET camera is limited. Situations where none or only one of the photons is detected (Figure 5.23(B)) are more common. We will refer to a situation when only one photon is detected during the coincidence as a *single detection* or simply a *single*.

Because of the nature of coincidence detection, as explained, some *coincidence detection errors* are introduced. There are two major effects that contribute to coincidence errors, namely *scatter coincidence errors* and *random coincidences.* Scatter coincidence error occurs when one of the annihilation photons undergoes the Compton scatter and is later detected by the camera (Figure 5.23(C)). The Compton scatter errors are easily correctable by discriminating the detected energy if PET detectors have perfect energy resolution. This would be done by not accepting detections with energies lower than 511 keV. However, the energy resolutions of the PET cameras are in the order of 20% making it impossible to perform accurate energy based discrimination. In fact, most of the PET cameras accept photons in the range of 350 keV to 650 keV (*energy window*) which results in a low discriminative power. Wide energy windows are used in order to decrease the chance of rejecting un-scattered 511 keV photons.

In Figure 5.23(C) the photons are scattered in the patient body. The scatter, however, can also occur outside the field of view. For example, scatter can occur on any object that is located in the room with the PET scanner. Scatter can also occur in the crystals of the PET scanner.

The other types of coincidence errors are due to random coincidences (Figure 5.23(D)). This occurs when photons which are detected in coincidence are products of annihilations of different positrons. In general, the number of random coincidences depends on the number of radioactive nuclei in the field of view of the camera and the width of the coincidence window at a time during which detections are considered as coincidences.

5.3.4.3   ROR for PET and TOF-PET

The basic premise of PET imaging is that if a coincidence detection in two detectors occurs, the origin of the actual decay is located on a line between the two detectors which is frequently termed in the literature as the *line of response* (LOR). The actual ROR differs quite substantially from the line. In the beginning of this section we explained that the actual decay location is different than annihilation and therefore the origin of the event (where decay occurred) is not exactly located on the LOR but rather a short distance away from it. The non-collinearity will also contribute to the deviation of the true origin location from the one predicted by the LOR.

Another significant source of deviation of actual ROR from the line modeled by LOR is the certainty of the location of the first interaction (LFI) of the 511 keV photon within the detector crystal (Figure 5.24). In this figure, each block has 4 crystals a1-a4 and b1-b4 for blocks a and b, respectively. In block a, the photon is absorbed in crystal a2 and the camera will likely assign the LFI to this crystal. For block b, the 511 keV photon is scattered in crystal b1 and then was absorbed somewhere on the border between crystals b2 and b3 (Figure 5.24). In such a case the LFI will likely be assigned erroneously to crystal b2 (average location of two interaction sites). The assigned LOR which is the line connecting crystals a2 and b2 is illustrated by the dotted line in Figure 5.24.

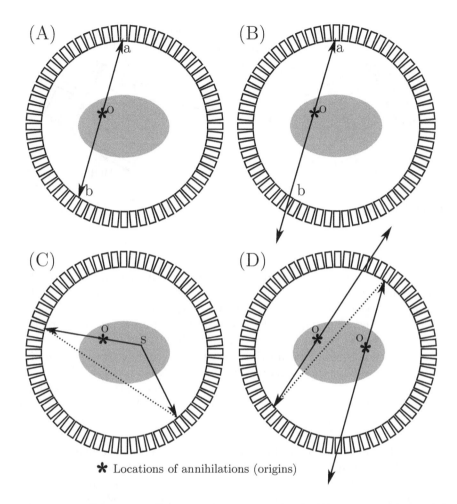

FIGURE 5.23 Schematics of various detections in PET. (A) true coincidence. (B) Single annihilation photon detection (a single). (C) Scatter coincidence detection (s = scatter location). (D) Random coincidence detection.

The LFI has two sources of uncertainty: (1) it is uncertain in which detector crystal LFI occurred, and (2) it is uncertain where within the crystal LFI occurred. Although it seems that these two sources of uncertainties can be summarized by simply saying that LFI within the detected block is uncertain, we differentiate these two sources of uncertainty because the decision about LFI is made at two different stages of data processing. In the first stage, the crystal in which the interaction occurred is determined based on signals from PMTs (in block detector). This is typically done during data pre-processing, even before the information about the count is injected into the list-mode file.

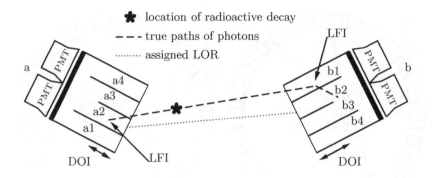

FIGURE 5.24   Detection of a direct (true) coincidence by two detector blocks (a and b).

In the other stage (typically during the data analysis/reconstruction) the approximation is made about the LFI within the volume of the crystal. For example, the assumption is made that LORs begin and end in the centers of the crystals. The inner surface of the crystal or average depth of interaction can also be used as beginnings and ends of LORs.

The uncertainty of unknown LFI is the cause of *parallax error*. We mention this term because it is frequently used in the literature but the parallax effects are only one type of errors that result from the uncertainty in LFI. Another frequently used term to describe some of the aspects of uncertain LFI is the *depth of interaction* (DOI), which is the depth measured from the surface of the crystal to the LFI.

On one hand uncertainty in LFI, positron range, and non-collinearity make the modeling of the exact ROR difficult. On the other hand the concept of the LOR seems too simplistic. In practical applications some compromise between accurate modeling and the need for simplicity for efficient numerical implementation has to be reached.

In PET each coincidence count contains additional information which is the difference in time of arrival of coincidence photons. With the current state-of-the-art hardware this difference can be measured with a precision of about 500 picoseconds. Therefore, assuming LOR model, this information pinpoints the location of the origin of the counts to a section on the LOR (see Figure 5.25) reducing substantially the intricacy of the ROR. This is shown in Figure 5.25. The LOR (line indicated by A—B in Figure 5.25 which is a simplified ROR) connects two crystals in which LFI was determined. It is assumed that the origin of annihilation photons is located on the line. For TOF-PET, only a section of the line C—D (thick black line) defines the possible locations of the origin of the annihilation. If voxel based volume representation is used, the shaded areas correspond to voxels that can contain the origin (Figure 5.25).

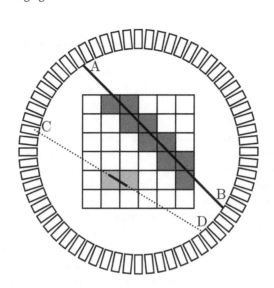

FIGURE 5.25   PET and TOF-PET count localization.

5.3.4.4   Quantitation of PET

Quantitation means that the actual values of radiodensities (numbers of radioactive nuclei per voxel) or values of some parameters that describe the underlying physiology are measured using PET. In order to achieve quantitation, several physical effects have to be taken into account. These effects are listed below. For detailed coverage of various methods used to measure and apply correction methods in order to achieve quantitation refer to Bailey et al. [5].

Attenuation Many of the photons that are emitted from the object will not reach the detectors because they are attenuated by the patient's body. The attenuated photons interact with the body mostly through Compton scatter and are deflected from the original path and their energy is reduced. These scattered photons may escape without further interaction with the body or camera, may interact again with the body, or escape the body and be detected by the camera. In either of those cases the event is considered to be attenuated. For large subjects, only a small percent of annihilations (both 511 keV photons) manage to escape the subject body. The attenuation is the most important factor that affects quantitation. When imaging small objects (small animal scanners) this effect is not as important. The information about attenuation is obtained from other imaging modalities (CT or MRI) and corrections for attenuation is incorporated in the system matrix (system matrix describe probabilities of events occurring in voxels and detected in detector elements) and effectively corrected for [5].

Normalization The relative positions of crystals are different and efficiencies of block detector vary. In addition, the PMT gains are not all exactly the same (and may change with time), and the scintillation crystals are not identical. This varying sensitivity to detection of coincidence photons needs to be taken into account in order to achieve quantitation in PET. The effect of different normalization is incorporated in the system matrix and effectively corrected for [5].

Compton scatter and randoms The coincidence detections of photons that underwent Compton scatter in the body (attenuation) or outside the body and random coincidences also affect the PET quantitation. The number of coincidence detections is artificially higher. This result in an overestimation of the radiodensity if corrective approaches are not used. There are various methods described in the literature [5] that can be used to correct for those effects. Compton scatter and randoms effects cannot be incorporated in the system matrix because they depend on the scanned object.

Dead time Dead-time effect occurs when there is a high radioactivity in the field of view. In such case, the detectors block can be hit with photons very frequently which will result in signal pile-ups. This subsequently will lead to misclassification of the detections as being outside the allowed energy window. It will result in losses of true counts. The dead-time correction methods are summarized in Bailey et al. [5]. Since the extent of dead-time is the function of the object, the correction for those effects cannot be done by the modification of the elements of the system matrix and different approaches must be used.

Even with corrections for all of the above effects, PET systems require a calibration in which objects with known radiodensity (activity) are scanned and then the radiodensity estimated from data is compared to the true value of radiodensity. The scaling factor is then established which is used to scale the voxel sensitivities in order to obtain the quantitative agreement between the true radiodensity and the radiodensity obtained by the PET scanner.

5.3.5 COMPTON IMAGING

We decided to include a short description of the Compton Imaging because of the different detection principle compared to SPECT or PET. Although the Compton Imaging in clinical and pre-clinical applications reached only the experimental states, it has been used in astronomy [108], physics, and security applications [65]. Similar to PET, the Compton Imaging is based on coincidence detections, but unlike PET only a single photon is considered. No collimation is used and instead the ROR is determined based on the localizations of the interactions within the detector and based on the deposited energies .

The schematics of basic operation of the Compton camera (CC) are presented in Figure 5.26. The gamma rays emitted in the object can be Compton

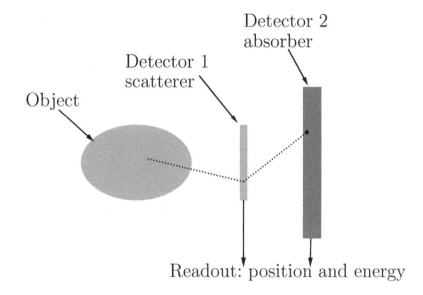

FIGURE 5.26 In Compton camera, gamma ray emitted in the object is Compton scattered on detector 1 (scatterer) and absorbed in detector 2 (absorber).

scattered in the first detector and be absorbed in the second detector. If interactions are registered in coincidence they are assumed to be caused by the same gamma photon. The energy deposited on at least one detector needs to be measured in order to be able to construct the ROR, that is, if the energy of the incoming gamma photons is known. In most applications the nuclide types being imaged are known and the energies of incoming gamma photons are known as well, but for other applications in astronomy or security the energies are not known or are ambiguous (several are possible). If the energies of incoming photons are not known the energies deposited on both detectors have to be measured.

The localization of the origins of the events in CC is based on measurements of the locations of the interactions using similar principles as for PET and SPECT. However, for CC the value of the energy deposited on the detectors is necessary as well. For PET or SPECT, the value of energy deposited in the detector by gamma photon is measured as well but only for the purpose of the discrimination between photons that were scattered or not in the object. For the CC (Figure 5.27) the surface of a half-cone is subtended which represents the possible locations of the origin of the event. This is based on the scatter and absorption location and scattering angle ψ (Figure 5.27). The scattering angle is determined based on the values of deposited energies.

Equation (5.2) shows the relationship between the energy of scattered pho-

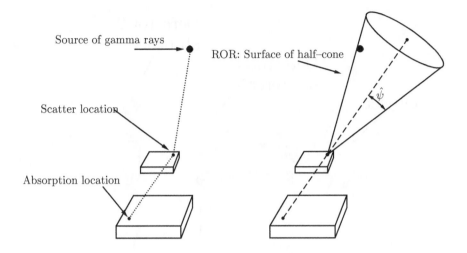

FIGURE 5.27   Construction of the ROR for Compton camera.

ton E_{sc}. Based on this equation the scattering angle ψ can be found using the values of energies deposited on scatterer (detector 1) E_1 and absorber (detector 2) E_2. We assume that energy of incoming photons is known and equal to E_0. Suppose further that E_1 and E_2 are measured by detectors 1 and 2 with the precisions ΔE_1 and ΔE_2. Based on those the estimate of $\hat{E}_{sc}$ is

$$\hat{E}_{sc} = \frac{(E_0 - E_1)(\Delta E_2)^2 + E_2(\Delta E_1)^2}{(\Delta E_1)^2 + (\Delta E_2)^2}$$

and the estimate of $\hat{\psi}$

$$\hat{\psi} = \arccos\left[1 - \frac{511 \text{ keV}}{E_0}\left(\frac{E_0}{\hat{E}_{sc}} - 1\right)\right].$$

The estimate of ψ can be used to construct the ROR as shown in Figure 5.27. The angle ψ can in theory vary between 0 and π radians. Since the surface of the half-cone ROR obtained by CC has two dimensions, the intricacy of such ROR is in the order of $0.66 \log_2 I$. Therefore the CC has much worse single-event localization capabilities compared PET or SPECT.

5.4   DYNAMIC IMAGING AND KINETIC MODELING

So far we have stated that the subject of the imaging is to determine the number of radioactive nuclei in the voxels of the imaged volume. If we assume that this number does not change over time, the analysis of this number would constitute the complete analysis of the problem. However, nuclear imaging techniques are inherently *functional methods* which means that they measure the function associated with physiological changes in the object. In many applications this function is measured by analysis of the changes of the distribution of the radioactive nuclei over time. The radioactive nuclides are attached to various biochemical compounds which undergo physiological processes that mimic the endogenous molecules in the subjects. We refer to those as radiotracers. Based on dynamics of radiotracer molecules the properties of endogenous molecules are inferred. The nuclear imaging that is used to monitor these changes over time will be referred to as *dynamic nuclear imaging* or *dynamic imaging*.

Typically the changes are characterized by using some models of the biological processes occurring in the region of interests (ROIs). The ROI is a group of voxels that represents some volume in the object. The complexity of biological processes exceeds by far the capabilities of the nuclear imaging to characterize fine details of the physiology. Therefore, simple but useful models of physiological processes are used. As the focus of this book is not on the physiology but the computational techniques, we refer the reader to detailed treatments of these topic to Berman et al. [9] and Morris et al. [73].

5.4.1 COMPARTMENTAL MODEL

One of the most popular models used for simulating of the dynamic behavior nuclear imaging is the *compartmental model*. In this model the radioactive tracers are assumed to be in one of several possible physiological/chemical states. To illustrate this, let's consider the following model:

Example 5.4: ^{18}FDG

The ^{18}FDG is a glucose analog ^{18}F 2-fluoro-2-deoxy-D-glucose. It is transported from blood plasma by facilitated diffusion into cells. Inside the cell, the ^{18}FDG is phosphorylated as is glucose. The difference between glucose and FDG metabolism is that phosphorylated FDG cannot be processed as regular glucose and remains in most of the cells. Using this as an example to illustrate different physiological states of the compound (^{18}FDG) that carries the nuclide, the three states can be identified: (1) ^{18}FDG is contained within the plasma, (2) ^{18}FDG is transported into cells and is present in the cells, and (3) the ^{18}FDG is phosphorylated and gets trapped in the cell. This is illustrated in Figure 5.28.

The physiological states of the tracer are called *compartments*. This is somewhat an unfortunate nomenclature. The term suggests that compartments are spatially disjoint. As described in the preceding example, the compartments (different physiological states of the compound) do not have to occupy spatially different volumes. This is illustrated in Figure 5.28.

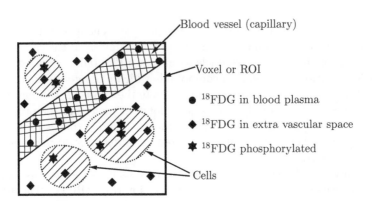

FIGURE 5.28 Representation of the voxel or ROI. The FDG tracer is assumed in either of three states: (1) blood plasma, (2) extra-vascular space, or (3) phosphorylated state.

Although the quantities of biochemical tracers are discrete and measured in the number of the tracer molecules present in the voxel or ROI, this number is usually high and the continuous approximations are used to simplify the modeling. The tracer concentrations are used and are referred by symbol $\mathbf{Q}(t)$. The $\mathbf{Q}(t)$ with elements $[Q_2(t), Q_3(t)...]$ has the size of the number of physiological compartments minus one. Symbol t indicates that Q depends on time. The indexing of the compartments in vector $\mathbf{Q}$ starts from 2 because the first compartment is reserved. We always identify compartment 1 as corresponding to the tracer in the blood plasma. The concentration of the tracer in compartment 1 (blood plasma) will normally be known and not be modeled. We will refer to this concentration as the *input function* and denote it as $C_P(t)$.

In the first approximation, the rate of tracer switching between compartments is a linear function of the concentration. The proportionality coefficients are called *kinetic parameters* and the symbol $k_{ii'}$ will be used to denote these coefficient. We adopt a convention that kinetic parameters are identified by two indices in this case i and i'. These indices indicate that the kinetic parameter describes the rate of switching to compartment i (first index) from compartment i' (second index).

Interpretation: The kinetic parameters can intuitively be interpreted as the chance per unit time that a tracer molecule changes the compartment.

An example of the three-compartment model is presented in Figure 5.29.

The differential equation governing the concentrations of a generalized kinetic model is

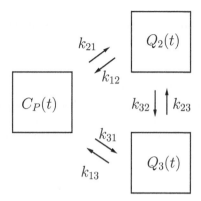

FIGURE 5.29 Three-compartment model. Compartment one corresponds to the tracer in the plasma ($C_P(t)$).

$$\frac{d\mathbf{Q}(t)}{dt} =$$

$$\begin{bmatrix} -\sum_{i=1}^{N} k_{i2} & k_{23} & k_{24} & \cdots & k_{2N} \\ k_{32} & -\sum_{i=1}^{N} k_{i3} & k_{34} & \cdots & k_{3N} \\ \vdots & \vdots & \vdots & \ddots & \vdots \\ k_{N2} & k_{N3} & k_{N4} & \cdots & -\sum_{i=1}^{N} k_{iN} \end{bmatrix} \mathbf{Q}(t) + \begin{bmatrix} k_{21} \\ k_{31} \\ \vdots \\ k_{N1} \end{bmatrix} C_P(t)$$

$$(5.13)$$

where N is the number of compartments including compartment 1. Assuming that at $t = 0$ the $\mathbf{Q}(t) = 0$ the above equation has a solution given by

$$\mathbf{Q}(t) = \exp(\mathbf{Q}(t)) * C_P(t) \begin{bmatrix} k_{21} \\ k_{31} \\ \vdots \\ k_{N1} \end{bmatrix}. \qquad (5.14)$$

The $*$ denotes the convolution. The exponent of a matrix can either be evaluated numerically or in a closed-form solution. The vector $[k_{21}, k_{31}, \ldots, k_{N1}]^T$ represented the vector of transfer rates from the input compartment $C_P(t)$ to other compartments.

5.4.2 DYNAMIC MEASUREMENTS

The most popular approach to dynamic imaging is summarized in Figure 5.30. The total acquisition time is divided into intervals or *time frames* and the data acquired in each time frame is considered (e.g., reconstructed) independently. It is possible to process the data using more sophisticated approaches where the correlation between time frames is taken into account or the division into time frames is not used at all [77], but in practice the approach based on the time-frames concept is used almost exclusively because of the ease of processing. Therefore, the measurement that is used in kinetic modeling represents the average concentration of the tracer in a voxel or ROI over some interval of time $[t_l, t_{l+1}]$ where l enumerates the time intervals and there are a total of L such intervals.

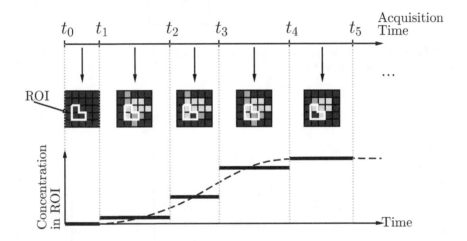

FIGURE 5.30 The acquisition time is divided into intervals. Data acquired during each interval is used to reconstruct images—one image per one interval. The ROI or multiple ROIs are defined and concentration values in the ROI are determined for each interval (time frame). The thick lines correspond to the average values of continuous function shown as a dashed line.

The average concentration in the time frame l is therefore expressed by

$$C_T(l) = (1 - f_v) \sum_{l=1}^{L} \int_{t_l}^{t_{l+1}} Q_l(t)dt + f_v \int_{t_l}^{t_{l+1}} C_P(t) \qquad (5.15)$$

The $C_T(l)$ indicates the average activity that is a sum of average activities of all compartments except the compartment one.

Notation: We use convention that if the function has an discrete argument it is a discrete function. For example, $C_T(l)$ indicated the discrete values. Therefore the $C_T(t)$ would indicate the continuous function of the concentration and is equal to

$$C_T(t) = (1 - f_v) \sum_{l=1}^{L} Q_l(t) + f_v C_P(t)$$

Another important observation in Equation (5.15) is that $C_T(l)$ is modeled as the sum of the average concentrations in the tissue compartments and the input function concentration (compartment 1). We define a tissue compartment as a compartment different than the blood. The coefficient f_v is interpreted as the ratio of the ROI that is spatially occupied by the blood (the input function).

5.5 APPLICATIONS OF NUCLEAR IMAGING

Nuclear imaging is mainly used in medicine for diagnosis and therapy monitoring of various diseases. Nuclear imaging is used for noninvasive studies of the fundamental molecular pathways inside the bodies in clinical and preclinical applications. In recent years the term *molecular imaging* was coined to describe imaging techniques used for visualization of molecular functions *in vivo*. Nuclear imaging (mainly SPECT and PET) are two key imaging modalities of molecular imaging. Other fields that use nuclear imaging are astronomy, environmental protection, and security, which are briefly discussed in Section 5.5.2.

5.5.1 CLINICAL APPLICATIONS

There is a large variety of uses for various radiotracers. The basic assumption is that the radiotracer should behave similar to or in a predictable manner compared to the substance that radiotracer mimics. The tracer is delivered to the patient and the idea is that it does not alter the physiological processes and therefore it should be delivered in trace amounts. There are exceptions to this rule as, for example, the radioactive water which is used to study the blood flow[3] in various parts of the body. Obviously since water constitutes the majority of the mass of biological organisms we do not have to be concerned that adding $H_2{}^{15}O$ alters the biochemical processes in the body. Measurements of the blood flow in the heart is one of the most popular clinical utilization of nuclear imaging. It is used with both SPECT (e.g., ^{99m}Tc-sestaMIBI) and PET (Ammonia ^{13}NH$_3$).

The most common PET radiotracer used today in clinics is the fluorodeoxyglucose or FDG (already mentioned in Example 5.4). It is an analog of glucose as the name indicates, but a slight chemical difference between

---

[3]The blood flow is defined as the amount of blood volume delivered per unit of the mass of the biological tissue per unit time.

FDG and glucose makes it an excellent imaging agent. The glucose or FDG is transported out of the blood vessels though facilitated diffusion.[4] After this glucose and FDG are phosphorylated. At this point, the physiological paths differ. The phosphorylated glucose is further metabolized through glycolysis. The phosphorylated FDG however is metabolized very slowly and in time scale corresponding to imaging time it can be assumed that phosphorylated FDG accumulates in the tissue (at least in most of organs). This remarkable property of the FDG makes is an excellent marker for measuring the speed glucose utilization in a simple manner. The concentration of accumulated phosphorylated FDG in tissue which is indicative of the glucose utilization was first demonstrated by Phelps et al. [80] and Sokoloff [99]. In clinical practice a measure called *standardized uptake value* (SUV) was developed which is the concentration of the FDG in some ROI divided by the total activity administered and mass of the body. Since most cancers have upregulated glucose metabolism, the SUV is used for making decisions regrading patient treatment planning and effectiveness.

Ligand-receptor studies are another popular application of the clinical nuclear imaging. In those applications *radioligands* are used which are biochemical compounds injected to blood circulation. Radioligands bind to *receptors* in the body and by measuring the concentration of the radioligands in time the affinity of the ligand to receptor can be studied using nuclear imaging techniques and dynamic imaging shortly introduced in Section 5.4.2. Examples of such application of nuclear imaging are studies that involve binding of various ligands to dopamine receptors in the brain and effects of this binding on schizophrenia patients [78].

In Table 5.4 examples of tracers used in nuclear imaging are presented with short summaries of applications.

5.5.2 OTHER APPLICATIONS

Translational and basic medical research: Another large area of application of nuclear imaging is the pre-clinical development of drugs and medical procedures. The ultimate goal of this development is to improve clinical tools used in medicine. This area of research is frequently called *translational research in medicine* to emphasize that the ultimate goal in research is to translate research findings to applications in humans. Following the *drug discovery* and before phase-0 (first-in-human trials) potential new drugs are evaluated for efficacy, toxicity, and pharmacokinetics. This is done *in vitro* and *in vivo* and preclinical nuclear imaging is an integral part of the *in vivo* evaluation. The preclinical imaging is done on various animals which to a lesser or higher

---

[4]Facilitated diffusion (also know as carrier-mediated diffusion) is a passive diffusion via special transport proteins which are embedded in the cell membrane.

TABLE 5.4
Examples of tracers used in nuclear imaging

Targeted molecular process	Radiotracer
Blood Flow:	
Diffusable (not trapped)	$H_2{}^{15}O$, ^{133}Xe, ^{99m}Tc-teboroxime (heart),
Diffusable (trapped)	$^{201}TcCl$ (heart), ^{82}Rb, ^{99m}Tc-tetrofosmin (heart), ^{62}Cu-PTSM[a], ^{99m}Tc-HMPAO[b] (brain),
Nondiffusable (trapped)	^{99m}Tc-albumin (lung)
Beta amyloid	^{18}F-PiB[c], ^{18}F-florbetapir, ^{18}F-flutemetamol
Blood volume	^{11}CO, ^{99m}Tc-red blood cells (RBC)
Lung ventilation	^{133}Xe, ^{81}Kr
Metabolism:	
Oxygen	$H_2{}^{15}O$
Oxidative	^{11}C-acetate
Glucose	^{18}F-fluorodeoxyglucose
Fatty acids	^{123}I-hexadecanoic acid, ^{11}C-palmitic acid
Osteoblastic activity	^{99m}Tc-methylene diphosphonate
Hypoxia	^{62}Cu-ATSM[d], ^{18}F-fluoromisonidazole
Proliferation	^{18}F-fluorothymidine
Protein synthesis	^{11}C-methionine, ^{11}C-leucine
Receptor-Ligand imaging:	
Dopaminergic	^{18}F-fluoro-L-DOPA[e], ^{11}C-raclopride, (^{18}F-fluoroethyl)spiperone
Benzodiazepine	^{18}F-flumazenil
Opiate	^{11}C-carfentanil
Serotonergic	^{11}C-altanserin
Adrenergic	^{123}I-metaiodobenzylguanidine (mIBG)
Somatastatin	^{111}In-octreotide
Estrogen	^{18}F-fluoroestradiol

[a]pyruvaldehyde bis(N^4-methylthiosemicarbazone)
[b]hexamethyl propylene amine oxime
[c]Pittsburgh compound B (analog of thioflavin T)
[d]diacetyl-bis(N^4-methylthiosemicarbazone)
[e]L-3,4-dihydroxyphenylalanine
Note: Table adopted from Cherry et al. [18].

degree are similar in their physiology to the physiology of humans. The nuclear imaging scanners are adopted for use with animal preclinical imaging and in general they are characterized by a smaller form factor to account for the smaller sizes of animals used in the translational research. The most frequently studied animals in pre-clinical are mice due to low cost, but also mice are quite different physiologically and typically used for first efficacy tests. Other animals used include rats, rabbits, pigs, dogs, or primates such as monkeys or apes.

The preclinical imaging can also be used for basic science investigations of the biological processes with no immediate goal of translating the findings to medicinal uses in humans [21]. In general, the dichotomy of basic and translational research is fluid because the basic research may contribute to medicine either by an increase in understanding of physiology or accidental findings (like the discovery of penicillin [42]). On the other hand, the translational research may contribute to growth of knowledge that does not directly apply in medicinal uses.

Astronomy: Nuclear imaging can be used in sub-MEeV/MeV γ-ray astronomy. The synthesis of radioactive nuclei occurs at the end-point of stellar evolution at which point stars of sufficient mass undergo massive explosions called *supernovae* and *novae*.[5] Radioactive nuclei and the products of the decays produced during this event are the direct tracers of the physical conditions of the super-explosions or novae [24]. Table 5.5 details some of the radio-activities observed in novae. Another significant source of gamma rays in the universe is the positron annihilation which creates 511 keV photons.

TABLE 5.5

Isotopes observed in novae (most apparent in the sky)

Isotope	Half-life (years)	Decay mode/daughter	Photon energies[a] (keV)
Sodium-22	2.6	β^+/Neon-22	1275
Aluminum-26	7.2×10^5	β^+/Magnesium-26	1809
Iron-60	2.6×10^6	β^-/Cobalt-60	59, 1173, 1332

[a] Main energy lines of the decay chain.
Note: Data from Takeda [104] and Hernanz [43].

Compton camera can be used to study gamma-ray astronomy. The basic concepts of the camera are explained in Section 5.3.5. The most successful implementation of the Compton camera was the Imaging Compton Telescope (COMPTELL) with two layers of gamma-ray detectors (scatterer and absorber) designed to operate for a range of 1 to 30 million electron volts (MeV). Another NASA-launched device is the Energetic Gamma Ray Experiment Telescope (EGRET) that uses Compton camera and operates on a much higher range of photon energies between 20 million (20 MeV) to 30 billion (30 GeV) electron volts (http://heasarc.gsfc.nasa.gov/docs/cgro/cgro/). The camera delivered to orbit in 1991 was removed from the orbit due to gyroscope failure in 1999.

---

[5]Supernovae and novae are different events; however, both are characterized by the creation of radioactive material.

The development of new detector materials lead to a design and construction of a new Compton imager that is planned to be delivered to orbit in 2015 as a part of ASTRO-H project. The Soft Gamma-ray Detector (SGD) [61, 105] combines a stack of Si strip detectors and CdTe pixel detectors to form a Compton telescope (see Figure 5.31). The Compton camera for this project is built from silicon strips (scatterer) and CdTe semiconductor detectors (absorber). The camera is shielded by using BGO crystal which is an active shield detecting escaping photons (Figure 5.31). Avalanche photodiodes (APDs) are used for amplification. The device is direction-sensitive by use of BGO collimator. Thin plate copper is used for collimation of low-energy gamma and x-rays (Figure 5.31).

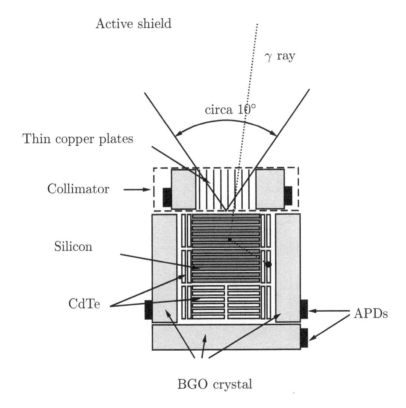

FIGURE 5.31 Conceptual drawing of soft-gamma detector using the Compton camera. Drawing made based on Tajima et al. [102].

Environmental and security applications: The clean-up efforts following environmental disaster like the one that occurred in Fukushima Daiichi nuclear power plant on March 11, 2011 require localization of radiation hotspots to ensure effective decontamination. In Fukushima ^{137}Cs, ^{134}Cs, and ^{131}I were released in large quantities. The nuclear imaging with portable Compton camera (weight of 1.5 kg) can be used to detect and characterize the activity in terms of the type of the isotope and the locations of the hotspots [56, 103].

Nuclear imaging can be used to detect sources of radiation and intercept stolen nuclear materials. In Lawrence Livermore laboratories, a Large Area Imager was developed that uses gamma-ray detection to pinpoint the source of radioactivity. The device consisted of a large area NaI crystal and collimators were placed on a track. The track can be driven around the danger area [114].

6 Statistical computing

In this chapter we provide details of implementations of Monte Carlo computational methods introduced in Chapter 4 to nuclear imaging discussed in Chapters 3 and 5. We apply the methods to nuclear imaging in which the imaged object is assumed to consist of voxels containing some number of radioactive nuclei. As described in Chapter 5 there is a large variety of scanner technologies geometries, principles of detection, etc. However, a common denominator between them is that if a decay occurs in a voxel of the object there is a constant deterministic chance specified by the system matrix that the event is detected in one of the detector elements of the camera. For example, the detector element can be defined as a coincidence detection between two crystals of either the PET camera or the Compton camera. The detector element can also be defined by the value of electric currents measured by the photomultiplier tubes in SPECT system. The currents are used to determine the location of the interaction of a photon in the crystal using Anger logic (Section 5.3.3.2). Based on the definition of the detector element the corresponding region of response (ROR) is defined which indicates the voxels where the detected event could originate. The above description is general and applies to any nuclear imaging system.

In addition to the common denominator described in the previous paragraph, different systems exhibit different characteristics and nuances. Since the goal of this book is to provide a general framework to application of Bayesian statistics in nuclear imaging we will not describe any specific implementations for specified modalities, but instead provide examples of the computation based on a very simple model. It is our intent to provide easy-to-understand examples of implementations of statistical principles that can be extended to more complex imaging systems. We see the examples included in this section as a guide for investigators or programmers wishing to implement the methods described in this book to specific problems they face. The analysis of any imaging system (with any complexity) with any physical characteristics can be implemented based on examples of Bayesian statistical computing included in this chapter.

6.1 COMPUTING USING POISSON-MULTINOMIAL DISTRIBUTION (PMD)

It should be clear that based on the model specified by Equations (3.34), (3.37) and (3.38) in Chapter 3 that the likelihood using PMD cannot be easily evaluated because it involves the sums of probabilities over the intersections of high-dimensional subsets and is not feasible from the computational point of view. However, as it will be demonstrated, sampling of the posterior of the quantity $\mathbf{y}$ (the complete data) is amazingly simple through the Origin Ensemble (OE) algorithm. The samples of the posterior of $\mathbf{y}$ can be trivially marginalized to provide samples of $\mathbf{c}$, $\mathbf{d}$, and $\mathbf{r}$ which will be shown in this chapter. There are several steps that lead to OE algorithm that are discussed in detail.

6.1.1   SAMPLING THE POSTERIOR

The basis of the main algorithm used for the analysis of the data is the general model of nuclear countind described by PMD derived in Chapter 3 and expressed by Equation (3.33) repeated here for convenience:

$$p(\mathbf{g}|\mathbf{c}) = \sum_{\mathbf{y} \in \mathbf{Y_c} \cap \mathbf{Y_g}} p(\mathbf{y}|\mathbf{c}). \tag{6.1}$$

The $\mathbf{g}$ is the measurement of the observable number of counts observed in detector elements. The size of this vector is equal to the number of detector elements K. The object is divided into I voxels and the number of events occurring in the voxels that were detected is denoted by vector $\mathbf{c}$. The vector $\mathbf{c}$ is an unobservable quantity (UQ) of the number of events emitted per voxel and detected. The element of vector $\mathbf{y}$ denoted as y_{ki} indicates the number of events that occurred in voxel i and were detected in detector element k.

The set of possible vectors $\mathbf{y}$ is denoted as $\mathbf{Y}$. Two subsets within $\mathbf{Y}$ are identified $\mathbf{Y_g} \subset \mathbf{Y}$ and $\mathbf{Y_c} \subset \mathbf{Y}$ such that $\mathbf{Y_g} = \{\mathbf{y} : \forall k \sum_{i=1}^{I} y_{ki} = g_k\}$ and $\mathbf{Y_c} = \{\mathbf{y} : \forall i \sum_{k=1}^{K} y_{ki} = c_i\}$. The $\mathbf{Y_c} \cap \mathbf{Y_g}$ indicates the intersection of these two subsets. The conditional $p(\mathbf{y}|\mathbf{c})$ was derived in Chapter 3 and was given by Equation (3.24).

$$p(\mathbf{y}|\mathbf{c}) = \prod_{i=1}^{I} p(\mathbf{y}_i|c_i) = \begin{cases} \prod_{i=1}^{I} \frac{c_i!}{(\epsilon_i)^{c_i}} \prod_{k=1}^{K} \frac{(\alpha_{ki})^{y_{ki}}}{y_{ki}!} & \text{for } \mathbf{y} \in \mathbf{Y_c} \\ 0 & \text{for } \mathbf{y} \notin \mathbf{Y_c} \end{cases} \tag{6.2}$$

where α_{ki} is the chance that if an event occurs in voxel i, it is detected in detector element k. The ϵ_i is the voxel sensitivity defined as $\epsilon_i = \sum_{k=1}^{K} \alpha_{ki}$ that expresses the chance that if an event occurs in voxel i it is detected by the camera (in any of the detector elements).

Using the statistical model expressed by Equation (6.1) and the Bayes Theorem the pre-posterior of $\mathbf{c}$ conditioned on observed quantity $\mathbf{g}$ is

$$p(\mathbf{c}|\mathbf{g}) = \frac{p(\mathbf{g}|\mathbf{c})p(\mathbf{c})}{p(\mathbf{g})} = \frac{1}{p(\mathbf{g})} \sum_{\mathbf{y} \in \mathbf{Y_c} \cap \mathbf{Y_g}} p(\mathbf{y}|\mathbf{c})p(\mathbf{c}) \tag{6.3}$$

The posterior seems quite difficult to compute considering that the sum runs over the intersection of two subsets and sampling from this intersection is not easy to implement numerically. However, in all examples of Bayesian analysis provided here we will compute the expectations of some quantities

$\phi(\mathbf{c})$ and as it will be shown next, those expectations are much easier to compute. We denote the expectation over the posterior as $E^{p(\mathbf{c}|\mathbf{g})}(.)$. Using Equation (6.3) the expectation is

$$E^{p(\mathbf{c}|\mathbf{g})}(\phi(\mathbf{c})) = \sum_{\mathbf{c}} \phi(\mathbf{c})p(\mathbf{c}|\mathbf{g}) = \frac{1}{p(\mathbf{g})} \sum_{\mathbf{c}} \phi(\mathbf{c}) \sum_{\mathbf{y}\in\mathbf{Y_c}\cap\mathbf{Y_g}} p(\mathbf{y}|\mathbf{c})p(\mathbf{c}) \quad (6.4)$$

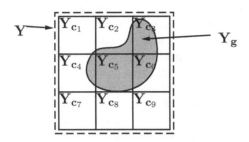

FIGURE 6.1   Illustration of disjoint subsets $\mathbf{Y_c}$. In this example the set $\mathbf{Y}$ is divided into nine disjoint subsets $\mathbf{Y_{c_1}}, \ldots, \mathbf{Y_{c_9}}$ that provide the *exact cover* of the set $\mathbf{Y}$.

Because subsets $\mathbf{Y_c}$ partition set $\mathbf{Y}$ which is illustrated in Figure 6.1 the $\sum_{\mathbf{c}}\sum_{\mathbf{y}\in\mathbf{Y_c}\cap\mathbf{Y_g}} = \sum_{\mathbf{y}\in\mathbf{Y_g}}$ and Equation (6.4) is simplified to

$$E^{p(\mathbf{c}|\mathbf{g})}(\phi(\mathbf{c})) = \frac{1}{p(\mathbf{g})} \sum_{\mathbf{y}\in\mathbf{Y_g}} \phi(\mathbf{c_y})p(\mathbf{y}|\mathbf{c_y})p(\mathbf{c_y}) \quad (6.5)$$

where $\mathbf{c_y}$ is implied by $\mathbf{y}$ such that $(c_y)_i = \sum_{k=1}^{K} y_{ki}$. The sum expressed by Equation (6.5) will be approximated using Markov Chain Monte Carlo (MCMC) methods described in Section 4.3. We create a chain of states $\mathbf{y}$ by application of the Metropolis–Hastings algorithm described in Section 4.4.4. The algorithm consists of two steps: (1) selection of new state $\mathbf{y}(s+1)$ given state $\mathbf{y}(s)$, and (2) acceptance of state $\mathbf{y}(s+1)$ given that the current state is $\mathbf{y}(s)$.

Notation: The $\mathbf{y}(s)$ indicates a vector $\mathbf{y}$ that corresponds to state s in the Markov chain of states. We can also think about the Markov state as an element of set $\mathbf{Y}$.

Important: By comparing Equation (6.4) and Equation (6.5) we conclude that samples of $\mathbf{y}$ from $\mathbf{Y_g}$ drawn with a chance proportional to $p(\mathbf{y}|\mathbf{c_y})p(\mathbf{c_y})$ are equivalent to samples of $\mathbf{c}$ from the posterior distribution $p(\mathbf{c}|\mathbf{g})$. These samples can be used either for computing the expectations as illustrated in Equation (6.5) or used to compute marginalized distributions (Section 4.1.2).

6.1.2 COMPUTATIONALLY EFFICIENT PRIORS

Notation: In the reminder of the book, we will indicate an element of $\mathbf{c_y}$ by simple $\mathbf{c}$. Therefore, we will keep in mind that $c_i = \sum_{k=1}^{K} y_{ki}$ always holds.

In order for the MCMC algorithm to be able to generate samples, some specific prior $p(\mathbf{c})$ in Equation (6.5) needs to be specified. There are two main considerations when designing the prior. First is the fact that prior should reflect our beliefs about the number of events per voxel, but at the same time the computational complexity that will result from use of the prior should be manageable. We introduce three separable and one non-separable priors. The separable priors are such that $p(\mathbf{c}) = p(c_1)p(c_2)\ldots p(c_I)$ and for non-separable priors this equation does not hold. For separable priors we will use (1) flat, (2) Gamma (3) Jeffreys prior and for non-separable prior we will use the entropy prior. We discuss the priors below:

Flat prior. The flat prior expresses beliefs that any number of emissions per voxel is equally probable and therefore

$$p_F(\mathbf{c}) \propto 1. \tag{6.6}$$

Using this prior

$$p(\mathbf{y}|\mathbf{c})p_F(\mathbf{c}) \propto \prod_{i=1}^{I} \frac{c_i!}{(\epsilon_i)^{c_i}} \prod_{k=1}^{K} \frac{(\alpha_{ki})^{y_{ki}}}{y_{ki}!} \tag{6.7}$$

Notation: To make the notation more compact we introduce $p(s) = p(\mathbf{y}|\mathbf{c})p(\mathbf{c})$ which we will refer to as the sampling distribution. This distribution is proportional to the probability of state s in Markov chain of states. We will use a subscript to indicate the type of prior that was used to derive the sampling distribution. For the case of the flat prior we have $p_F(s) = p(\mathbf{y}|\mathbf{c})p_F(\mathbf{c})$.

Therfore,

$$p_F(s) = p(\mathbf{y}|\mathbf{c})p_F(\mathbf{c}) \propto \prod_{i=1}^{I} \frac{c_i!}{(\epsilon_i)^{c_i}} \prod_{k=1}^{K} \frac{(\alpha_{ki})^{y_{ki}}}{y_{ki}!}. \tag{6.8}$$

Gamma prior. As detailed in Chapter 3 statistics of the number of emissions of events from voxels can be approximated by the Poisson distribution. Suppose that for the purpose of postulating prior we assume the prior knowledge about expected number of events that will occur per voxel ϑ_i (as defined in Section 3.4) and that ϑ_i is known with a confidence expressed by the parameter ω_i. We use *gamma distribution* to express those beliefs [95, 96] as

$$p_G(\mathbf{f}) = \prod_{i=1}^{I} p(f_i) = \frac{1}{Z} \prod_{i=1}^{I} (f_i)^{\vartheta_i \omega_i - 1} e^{-\omega_i f_i}. \tag{6.9}$$

By Z the normalization constant is indicated. We always use symbol Z to indicate the constant, but its value will vary depending on the context. The ϑ_i and ω_i are such that $\vartheta_i \omega_i \geq 1$. This guarantees that the integral of the distribution is finite and can be normalized to 1. For large values of parameter ω_i (rate parameter of gamma distribution) the prior distribution is strongly peaked around ϑ_i which indicates the high confidence and conversely the low value of ω_i indicates the low confidence. In the limit of $\omega \to 0$ the prior approximates the flat prior.

With this definition (Equation (6.9)), the gamma prior of $\mathbf{c}$ (number of event per voxel that occurred during the experiment) can be computed as

$$
p_G(\mathbf{c}) = \int_{\mathbf{f}} p(\mathbf{c}|\mathbf{f}) p(\mathbf{f}) = \int_0^\infty df_1 p(c_1|f_1) p(f_1) \times \ldots \times \int_0^\infty df_I p(c_I|f_I) p(f_I) \tag{6.10}
$$

where $p(c_i|f_i)$ is Poisson distributed

$$
p(c_i|f_i) = \frac{(f_i \epsilon_i)^{c_i} e^{-f_i \epsilon_i}}{c_i!} \tag{6.11}
$$

The above distribution can be derived from the Poisson distribution of $p(d_i|f_i)$ and binomial distribution of $p(c_i|d_i)$. This derivation is shown in Appendix E. The integral expressed by Equation (6.10) after arrangement of terms and using Equations (6.9) and (6.11) is

$$
p_G(\mathbf{c}) = \frac{1}{Z} \prod_{i=1}^I \frac{(\epsilon_i)^{c_i}}{c_i!} \int_0^\infty df_i (f_i)^{c_i + \vartheta_i \omega_i - 1} e^{-f_i(\epsilon_i + \omega_i)}. \tag{6.12}
$$

The integral $\int_0^\infty dx x^a e^{-bx} = \Gamma(a+1)/b^a$ where $\Gamma(x)$ is the gamma function and the above evaluates to

$$
p_G(\mathbf{c}) = \frac{1}{Z} \prod_{i=1}^I \frac{\Gamma(c_i + \vartheta_i \omega_i)(\epsilon_i)^{c_i}}{c_i!(\epsilon_i + \omega_i)^{c_i + \vartheta_i \omega_i}} = \frac{1}{Z} \prod_{i=1}^I \frac{\Gamma(c_i + \vartheta_i \omega_i)(\epsilon_i)^{c_i}}{c_i!(\epsilon_i + \omega_i)^{c_i}} \tag{6.13}
$$

where we pull the constant the term $(\epsilon_i + \omega_i)^{\vartheta_i \omega_i}$ to the constant Z. The final form of the sampling distribution with the gamma prior is therefore

$$
p_G(s) = p(\mathbf{y}|\mathbf{c}) p_G(\mathbf{c}) = \frac{1}{Z} \prod_{i=1}^I \frac{\Gamma(c_i + \vartheta_i \omega_i)}{(\epsilon_i + \omega_i)^{c_i}} \prod_{k=1}^K \frac{(\alpha_{ki})^{y_{ki}}}{y_{ki}!} \tag{6.14}
$$

Jeffreys prior: In order to use Jeffreys prior we will again assume for the sake of the derivation, each voxel i is expected to generate f_i events through the Poisson process during the scan. The idea of Jeffreys prior is that it is proportional to the square root of the determinant of the *Fisher information* matrix [66]. We do not discuss in detail the Fisher information or properties of the Jeffreys prior such invariance under reparametrization and an interested reader should consult the classical works of Jeffreys [49] and Lehmann and Casella [66].

Since the assumption of f_i for each voxel implies the Poisson distribution, the Fisher information for 1D Poisson process is equal to $(f_i)^{-1}$. It follows that the Jeffreys prior is the square root of this value and equal to

$$p_J(\mathbf{f}) \propto \prod_{i=1}^{I} (f_i)^{-\frac{1}{2}} \tag{6.15}$$

The proportionality sign is used because the prior is improper and the distribution of the prior cannot be normalized to 1. Again using Poisson distribution (Equation (6.11)) and marginalizing f_i's we obtain prior $p_J(\mathbf{c})$ as

$$p_J(\mathbf{c}) = \int_{\mathbf{f}} p(\mathbf{c}|\mathbf{f})p(\mathbf{f}) \propto \prod_{i=1}^{I} \frac{1}{c_i!} \int_0^{\infty} df_i e^{-f_i} f_i^{c_i - \frac{1}{2}} = \prod_{i=1}^{I} \frac{\Gamma(c_i + \frac{1}{2})}{c_i!} \tag{6.16}$$

Using the Jeffreys prior defined by Equation (6.16) the sampling distribution is

$$p_J(s) = p(\mathbf{y}|\mathbf{c})p_G(\mathbf{c}) \propto \prod_{i=1}^{I} \frac{\Gamma(c_i + \frac{1}{2})}{(\epsilon_i)^{c_i}} \prod_{k=1}^{K} \frac{(\alpha_{ki})^{y_{ki}}}{y_{ki}!} \tag{6.17}$$

Entropy prior. The entropy prior is an example of non-separable prior that can be efficiently implemented. The idea of this prior comes from the statistical physics. Suppose we consider a vector $\mathbf{c}$ as the macroscopic state. The arrangement of events between I voxels constitutes the microscopic state and therefore there are many microscopic states that correspond to a single macroscopic state. This concept is illustrated in Figure 6.2.

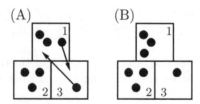

FIGURE 6.2 Representation of macroscopic and microscopic states for 3-voxel system.

Assuming that each microscopic state has the same probability[1] we have that for a given $\mathbf{c}$, the probability of the macrostate $\mathbf{c}$ is defined as

---

[1]This is the basic assumption used in statistical physics.

$$p(\mathbf{c}) \propto (\# \text{ of microstates corresponding to } \mathbf{c})^{\beta} \qquad (6.18)$$

The idea of microscopic and macroscopic states is explained in Figure 6.2. In Figure 6.2(A) the macroscopic state is defined by $\mathbf{c} = [3,3,1]$. If one of the events from voxel 1 is moved to voxel 3 and the event from voxel 3 is moved to voxel 1 the macroscopic state pictured in (B) is obtained. Since the number of events in voxels is the same as in Figure 6.2(A) it is the same macroscopic state. However, (A) and (B) correspond to two different microscopic states.

The constant β^2 is such that it can have values between 0 and 1 and corresponds to the strength of the belief that the probability of $\mathbf{c}$ is proportional to the number of microstates. If β is 1 the probability is directly proportional to the number of microstates and if $\beta = 0$ the prior assigns equal probability to all macrostates and therefore is equivalent to the flat prior.

From combinatorics we have the number of ways a number of events can be distributed amount I voxels to obtain vector $\mathbf{c}$ is

$$(\# \text{ of microstates corresponding to } \mathbf{c}) = \frac{\left(\sum_{i=1}^{I} c_i\right)!}{\prod_{i=1}^{I} c_i!}. \qquad (6.19)$$

Using Equations (6.18) and (6.19) we obtain what we refer to as the *entropy prior* as

$$p_E(\mathbf{c}) = \frac{1}{Z}\left(\frac{(\sum_{i=1}^{I} c_i)!}{\prod_{i=1}^{I} c_i!}\right)^{\beta} \qquad (6.20)$$

The sampling distribution with entropy prior is therefore

$$p(\mathbf{y}|\mathbf{c})p_E(\mathbf{c}) = \frac{1}{Z}\left(\frac{(\sum_{i=1}^{I} c_i)!}{\prod_{i=1}^{I} c_i!}\right)^{\beta} \prod_{i=1}^{I} \frac{c_i!}{(\epsilon_i)^{c_i}} \prod_{k=1}^{K} \frac{(\alpha_{ki})^{y_{ki}}}{y_{ki}!} \qquad (6.21)$$

The term in the numerator $\sum_{i=1}^{I} c_i$ is constant and equal to the total number of detected counts. Therefore, it can be pulled inside the constant Z. Taking this into account Equation (6.21) becomes

$$p_E(s) = p(\mathbf{y}|\mathbf{c})p_E(\mathbf{c}) = \frac{1}{Z}\prod_{i=1}^{I} \frac{(c_i!)^{1-\beta}}{(\epsilon_i)^{c_i}} \prod_{k=1}^{K} \frac{(\alpha_{ki})^{y_{ki}}}{y_{ki}!} \qquad (6.22)$$

---

[2]In correspondence to statistical physics β can be interpreted as proportional to inverse of temperature in microcanonical ensemble [33] in Equation (6.18), and in such a case the range of value of β would be from 0 to ∞.

6.1.3   GENERATION OF MARKOV CHAIN

All four sampling distributions corresponding to different priors expressed by Equations (6.8), (6.14), (6.17) and (6.22) are quite similar. The priors were designed so that the sampling distributions have this similar form for which the efficient Markov chain sampling algorithm can be specified.

The goal of this section is to build an algorithm that will create a chain as shown in Figure 6.3(A). We will always start with some initial $\mathbf{y}(0)$. We then append the chain with new states where a state $\mathbf{y}(s + 1)$ will depend only on the preceding state $\mathbf{y}(s)$. In our implementation the subsequent states in the Markov chain of states may differ only by location of a single origin or be identical. For 3-voxel 6-event model shown in Figure 6.3(B) the two states (s and $s + 1$) differ only by location of single event. For state s the event origin is in voxel 1 and for $s+1$ it is in voxel 3. Figure 6.3(C) shows two states s and $s + 1$ that are identical. As we recall from Chapter 4 it is acceptable to have identical states in the Markov chain as long as the algorithm that generates the states obey conditions required.

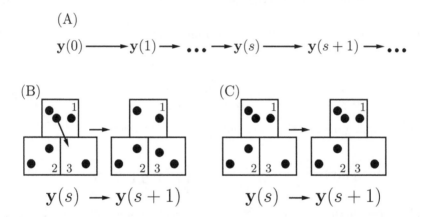

FIGURE 6.3   (A) The Markov chain built from $\mathbf{y}$ vectors. (B) and (C) Markov moves for 3-voxel 6-event model.

Before we provide the algorithm for generation of these states we derive the ratios of sampling distributions between states that differ only by the location of the origin of one event. Suppose that we consider two states $\mathbf{y}(s)$ and $\mathbf{y}(s + 1)$. We design the origin ensemble (OE) algorithm such that the neighboring states in the chain may (1) differ by a location of an origin of one event (Figure 6.3(B)) or (2) be identical (Figure 6.3(C)). For an imaging system with I voxels and K detector elements, without loss of generality, we assume states s and $s + 1$ may differ by the location of the origin of a count detected in detector element k. The voxel that the event is located in state s is i and may be located state $s + 1$ in voxel i'. The ratios of sampling distributions of the state $s + 1$ and s can be found from Equations (6.8), (6.14), (6.17) and (6.22). The details of derivation of those ratios are provided in Appendix D. Here we provide only the results of the derivations which are amazingly simple. This simplicity is the key for the computational efficiency of the algorithm.

$$\frac{p_F(s+1)}{p_F(s)} = \frac{\alpha_{ki'}\epsilon_i(c_{i'}+1)y_{ki}}{\alpha_{ki}\epsilon_{i'}c_i(y_{ki'}+1)} \tag{6.23}$$

$$\frac{p_G(s+1)}{p_G(s)} = \frac{\alpha_{ki'}(\epsilon_i+\omega_i)(c_{i'}+\vartheta_{i'}\omega_{i'})y_{ki}}{\alpha_{ki}(\epsilon_{i'}+\omega_{i'})(c_i+\vartheta_i\omega_i-1)(y_{ki'}+1)} \tag{6.24}$$

$$\frac{p_J(s+1)}{p_J(s)} = \frac{\alpha_{ki'}\epsilon_i\left(c_{i'}+\frac{1}{2}\right)y_{ki}}{\alpha_{ki}\epsilon_{i'}\left(c_i-\frac{1}{2}\right)(y_{ki'}+1)} \tag{6.25}$$

$$\frac{p_E(s+1)}{p_E(s)} = \frac{\alpha_{ki'}\epsilon_iy_{ki}}{\alpha_{ki}\epsilon_{i'}(y_{ki'}+1)}\left(\frac{c_{i'}+1}{c_i}\right)^\beta \tag{6.26}$$

6.1.4   METROPOLIS–HASTINGS ALGORITHM

The Metropolis–Hastings algorithm was discussed in Section 4.4.4. The idea of this algorithm is that once the new state $s + 1$ in the Markov chain is selected it can be stochastically accepted as a new state with some acceptance probability. If the new state is not accepted the new state $s + 1$ is unchanged and is the same as s. The acceptance ratio of the Metropolis–Hastings algorithm was defined as x in Equation (4.52) as

$$x = \frac{p(s+1)S(s+1 \to s)}{p(s)S(s \to s+1)} \tag{6.27}$$

where $S(s \to s+1)$ and $S(s+1 \to s)$ are selection probabilities. That is, the probabilities that if the chain is in state s the next selected state is $s + 1$ and vice versa.

Examining Equation (6.27) it is noted that the ratios $p(s + 1)/p(s)$ are defined by Equations (6.23)–(6.26). We are missing the critical part of the algorithm. The algorithm for the selection of state $s + 1$ when chain is in the state s is missing. When $S(s \to s + 1)$ and $S(s + 1 \to s)$ are defined, the Metropolis-Hastings algorithm will be fully defined as well.

As described earlier in this section the states s and $s + 1$ differ by location of origin of only one event. The selection algorithm that is specified next will define how this event is chosen and how the new potential location for the event is designated. This algorithm is quite simple and consists of two steps:

Step 1 Randomly select one of the detected events. During the run of the algorithm all detected events have to be catalogued. The information about which detector element the event was detected as well as the voxel where the event origin is located for the current state of Markov chain is known for each event. Therefore, once the event is randomly selected, the detector element k and voxel i are known from the catalog.

Step 2 Randomly select a new voxel for the event. There are two approaches to doing so. Approach 1 should be preferable for computational efficiency but depending on the problem at hand and difficulties in implementing this approach approach 2 can be used. In approach 1 (preferable approach) we choose a new voxel i' as a candidate that contains the origin of the event selected in step 1 with the chance proportional to $\alpha_{ki'}/\sum_{j=1}^{I}\alpha_{kj}$. In approach 2 we select one of the voxels and each voxel is selected with the same chance equal to $1/I$ (I is the number of voxels).

Using the above description the $S_1(s \to s+1)$ where subscript 1 stands for the approach 1 is

$$S_1(s \to s+1) = \frac{y_{ki}}{N} \frac{\alpha_{ki'}}{\sum_{j=1}^{I}\alpha_{kj}} \tag{6.28}$$

The first term in Equation (6.28) is equal to the chance that one of the events detected in k with the locations of origin in voxel i is selected in step 1. Similarly the

$$S_2(s \to s+1) = \frac{y_{ki}}{N}\frac{1}{I} \tag{6.29}$$

Using similar considerations the $S_1(s+1 \to s)$ and $S_2(s+1 \to s)$ are specified as:

$$S_1(s+1 \to s) = \frac{y_{ki'}+1}{N} \frac{\alpha_{ki}}{\sum_{j=1}^{I}\alpha_{kj}} \tag{6.30}$$

$$S_2(s+1 \to s) = \frac{y_{ki'}+1}{N}\frac{1}{I} \tag{6.31}$$

There is the $+1$ in the numerator because the system is considered in Markov state after successful "move," so from definition the number of event origins in voxel i' that were detected in k is $y_{ki'}+1$ as the value $y_{ki'}$ is defined for the state s. It is illustrated in Figure 6.4.

Using the ratios of sampling statistics defined by Equations (6.23)–(6.26) and selection probabilities specified in Equations (6.28)–(6.31) the Metropolis–Hasting acceptance ratio x (Equation (6.27)) can be fully specified. There are eight different acceptance ratios x depending on the prior and the algorithm used for selection of new voxels during Markov moves. For example, the x_{F1} indicates the Metropolis–Hastings acceptance threshold for the sampling distributions obtained using the flat prior and the selection algorithm 1.

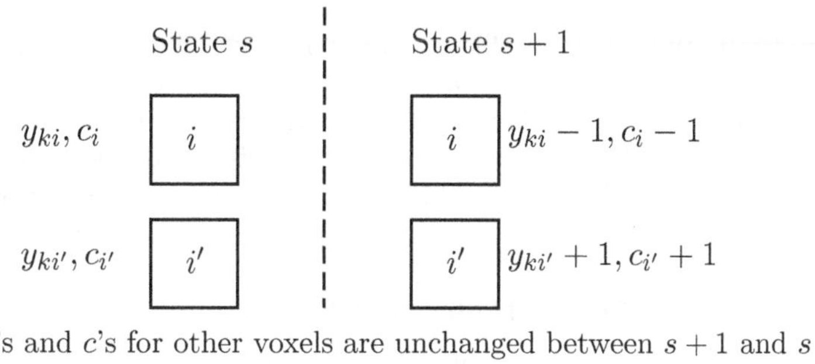
State s
State $s+1$
y_{ki}, c_i
i
i
$y_{ki}-1, c_i-1$
$y_{ki'}, c_{i'}$
i'
i'
$y_{ki'}+1, c_{i'}+1$
y's and c's for other voxels are unchanged between $s+1$ and s

FIGURE 6.4 Two subsequent states in the OE algorithm Markov chain if the "move" is successful. Squares represent voxels.

List of Metropolis-Hastings acceptance ratios

Flat prior:

$$x_{F1} = \frac{(c_{i'}+1)\epsilon_i}{c_i \epsilon_{i'}} \tag{6.32}$$

$$x_{F2} = \frac{\alpha_{ki'}(c_{i'}+1)\epsilon_i}{\alpha_{ki} c_i \epsilon_{i'}} \tag{6.33}$$

Gamma prior:

$$x_{G1} = \frac{(c_{i'}+\vartheta_{i'}\omega_{i'})(\epsilon_i+\omega_i)}{(c_i+\vartheta_i\omega_i-1)(\epsilon_{i'}+\omega_{i'})} \tag{6.34}$$

$$x_{G2} = \frac{\alpha_{ki'}(c_{i'}+\vartheta_{i'}\omega_{i'})(\epsilon_i+\omega_i)}{\alpha_{ki'}(c_i+\vartheta_i\omega_i-1)(\epsilon_{i'}+\omega_{i'})} \tag{6.35}$$

Jeffreys prior:

$$x_{J1} = \frac{\left(c_{i'}+\frac{1}{2}\right)\epsilon_i}{\left(c_i-\frac{1}{2}\right)\epsilon_{i'}} \tag{6.36}$$

$$x_{J2} = \frac{\alpha_{ki'}\left(c_{i'}+\frac{1}{2}\right)\epsilon_i}{\alpha_{ki}\left(c_i-\frac{1}{2}\right)\epsilon_{i'}} \tag{6.37}$$

Entropy prior:

$$x_{E1} = \frac{\epsilon_i}{\epsilon_{i'}}\left(\frac{c_{i'}+1}{c_i}\right)^{1-\beta} \tag{6.38}$$

$$x_{E2} = \frac{\alpha_{ki'}\epsilon_i}{\alpha_{ki}\epsilon_{i'}}\left(\frac{c_{i'}+1}{c_i}\right)^{1-\beta} \tag{6.39}$$

The above ratios are simple which is the result of the design of the Markov chain and allow an efficient numerical implementation. For the selection algorithm 1, the system matrix α does not have to be explicitly used for the calculation of the Metropolis–Hastings acceptance ratio. It is an important feature of this algorithm and may simplify modeling of many complex imaging systems. It is assumed however that for this method it is possible to select voxels proportionally to their value of the system matrix.[3]

6.1.5 ORIGIN ENSEMBLE ALGORITHMS

At this point we have all components to specify the origin ensemble (OE) algorithm[4] which is a MCMC algorithm for sampling the Poisson-multinomial posterior of the number of emissions per voxels that were detected described by **c** conditioned on acquired data **g**. All steps of the OE algorithm were already discussed in previous sections and are given below:

> **Origin ensemble algorithm for sampling $p(\mathbf{c}|\mathbf{g})$:**
>
> 1. Select starting state $\mathbf{y}(0)$. This can be done by randomly selecting for all counts in each detector element k voxels i with nonzero α_{ki}. This process may be considered as a guess about the locations of origins of the detected events.
> 2. Randomly select one of the detected counts. Note the detector element k that it was detected in and note the voxel i in which the origin of this event is located.
> 3. Randomly select a candidate new voxel i' for the event origin (selected in step 2). This can be done by either of the two algorithms described in Section 6.1.4.
> 4. Move the origin of the event selected in step 2 to the candidate voxel i' with a chance equal to
> $$\min{(1, x)}$$
> where the x is the acceptance ratio defined by one of the Equations (6.32)–(6.39). If the move is unsuccessful, the origin of the event remains assigned to voxel i.
> 5. Repeat the last three steps. The last three steps constitute a generation of a new Markov state in the Markov chain of states of the OE algorithm.

---

[3]This is the case for PET and TOF-PET imaging because the new location can be randomly selected on a LOR (or appropriately weighted LOR for TOF-PET) and then the voxel containing the location can easily be determined. Such algorithm implicitly simulates voxel selection proportional to the values of elements of the system matrix.

[4]The algorithm was first given in Sitek [94]. The derivation of the algorithm in this publication used *ad hoc* arguments. The statistical theory behind the algorithm was first shown by Sitek [95].

We define an iteration of this algorithm as steps 2 through 4 repeated N times (N is the total number of detected counts).

The above algorithm provides samples of the posterior of $\mathbf{c}$. However, the quantity $\mathbf{c}$ is typically not of direct interest because the number of emissions per voxel that are detected is affected by photon attenuation and non-uniform sensitivity of voxels throughout the images volume. We are interested in quantities such as $\mathbf{d}$ or $\mathbf{r}$ defined in Chapters 3 and 5 which correspond to the total number of events (decays detected or not) per voxel and the total number of original radioactive nuclides per voxel at the beginning of the scan. These quantities provide information about the spacial distribution of the tracer in scanned subjects.

Therefore the posterior of these quantities is needed. Obtaining these posteriors can be done quite easily because the statistical models of the data based on those quantities are defined in Chapter 3 by Equations (3.37) and (3.38). Because of the conditional independence of $\mathbf{d}$ and $\mathbf{g}$ given $\mathbf{c}$ and the conditional independence of $\mathbf{r}$ and $\mathbf{g}$ given $\mathbf{c}$ obtaining samples from posteriors $p(\mathbf{d}|\mathbf{g})$ and $p(\mathbf{r}|\mathbf{g})$ can be achieved by a straightforward extension of the OE algorithm already derived for sampling the $p(\mathbf{c}|\mathbf{g})$ posterior. It is the most efficient approach to do so. We only provide the derivation for OE sampling of $p(\mathbf{d}|\mathbf{g})$ but the derivation of the sampler of the posterior $p(\mathbf{r}|\mathbf{g})$ follows almost identical steps.

Suppose that we would like to evaluate the following expectation:

$$E^{p(\mathbf{d}|\mathbf{g})}(\boldsymbol{v}(\mathbf{d})) = \sum_{\mathbf{d}} \boldsymbol{v}(\mathbf{d})p(\mathbf{d}|\mathbf{g}) \tag{6.40}$$

The expectation looks very similar to previously considered expectation over the posterior of $\mathbf{c}$ given in Equation (6.4). The only difference is that it is done over the posterior of $\mathbf{d}$. Because of the conditional statistical independence of $\mathbf{d}$ and $\mathbf{g}$ given $\mathbf{c}$ shown in Chapter 3 we have that

$$p(\mathbf{d}|\mathbf{g}) = \sum_{\mathbf{c}} p(\mathbf{d}|\mathbf{c})p(\mathbf{c}|\mathbf{g}). \tag{6.41}$$

Using Equations (6.40) and (6.41) and interchanging order of summations we obtain

$$E^{p(\mathbf{d}|\mathbf{g})}(\boldsymbol{v}(\mathbf{d})) = \sum_{\mathbf{c}}\sum_{\mathbf{d}} \boldsymbol{v}(\mathbf{d})p(\mathbf{d}|\mathbf{c})p(\mathbf{c}|\mathbf{g}) =$$

$$\sum_{\mathbf{c}} E^{p(\mathbf{d}|\mathbf{c})}(v(\mathbf{d}))p(\mathbf{c}|\mathbf{g}) = \sum_{\mathbf{c}} \phi(\mathbf{c})p(\mathbf{c}|\mathbf{g}) \tag{6.42}$$

where by $\phi(\mathbf{c}) = E^{p(\mathbf{d}|\mathbf{c})}(\boldsymbol{v}(\mathbf{d}))$ we defined the expectation of some function of $\boldsymbol{v}$ of $\mathbf{d}$ $\boldsymbol{v}(\mathbf{d})$ over the distribution $p(\mathbf{d}|\mathbf{c})$. We will elaborate on this expectation shortly, but first we note that based on Equation (6.42) the sampling from the posterior $p(\mathbf{d}|\mathbf{g})$ (see Equation (6.40)) of $\boldsymbol{v}(\mathbf{d})$ is equivalent to sampling from $p(\mathbf{c}|\mathbf{g})$ (which we already know how to do) of $\phi(\mathbf{c})$ defined as $E^{p(\mathbf{d}|\mathbf{c})}(\boldsymbol{v}(\mathbf{d}))$.

Let's have a closer look at the $E^{p(\mathbf{d}|\mathbf{c})}(v(\mathbf{d}))$. The computation of this expectation is much easier than the expectation over $p(\mathbf{c}|\mathbf{g})$ and the direct sampling algorithm (no Markov chain is necessary) is available. To demonstrate this, consider that number of events that occur per voxel which are independent and therefore

$$p(\mathbf{d}|\mathbf{c}) = \prod_{i=1}^{I} p(d_i|c_i) \tag{6.43}$$

Thus, to generate an independent sample of $\mathbf{d}$ from $p(\mathbf{d}|\mathbf{c})$ it is sufficient to generate I samples from 1D distributions $p(d_i|c_i)$. At this point the only missing element is a method for generation of samples from $p(d_i|c_i)$. The d_i is the number of events that occurred in voxel i. These events may or may not be detected and the probability that they are detected in any of the detector elements is the same for each of the events and equal to the voxel sensitivity ϵ_i. Therefore, the d_i can be considered as the number of trials and $d_i - c_i$ the number of successes in Bernoulli trial with probability of success equal to $1 - \epsilon_i$ and c_i number of failures which occur with probability ϵ_i. We use the awkward terminology here naming the event that is not detected a *success* and the event that is detected as a *failure* because this will let us introduce in a straightforward way the distribution that governs $p(d_i|c_i)$.

If the number of failures in Bernoulli trials is known and it is larger than 0, then the distribution of the number of successes follows the *negative binomial distribution*[5] $\mathcal{NB}(a|b,c)$ where a is the number of successes, b is the known number of failures, and c is the known probability of success. Using these definitions it is now easy to specify

$$p(d_i|c_i) = \mathcal{NB}(d_i - c_i|c_i, 1 - \epsilon_i) = \binom{d_i - 1}{d_i - c_i}(\epsilon_i)^{c_i}(1 - \epsilon_i)^{d_i - c_i}. \tag{6.44}$$

The above makes the direct numerical draws from $p(\mathbf{d}|\mathbf{c})$ almost effortless as there are many numerical algorithms available for direct draws from the negative binomial distribution [82]. Using the above considerations, the OE algorithm for sampling from $p(\mathbf{d}|\mathbf{g})$ is specified next. This algorithm differs from the original OE given on page 190 by the addition of step 5.

Origin ensemble algorithm for sampling $p(\mathbf{d}|\mathbf{g})$:

1. Select starting state $\mathbf{y}(0)$. This can be done by randomly selecting for all counts in each detector element k voxels i with nonzero α_{ki}. This process may be considered as a guess about the locations of origins of the detected events.

---

[5]The distribution of the number of successes when the number of trials is known for Bernoulli trials is governed by the binomial distribution discussed in Chapter 3.

2. Randomly select one of the detected counts. Note the detector element k that it was detected in and note the voxel i in which the origin of this event is located.

3. Randomly select a candidate new voxel i' for the event origin (selected in step 2). This can be done by either of the two algorithms described in Section 6.1.4.

4. Move the origin of the event selected in step 2 to the candidate voxel i' with a chance equal to

$$\min(1, x)$$

where the x is the acceptance ratio defined by one of the Equations (6.32)–(6.39). If the move is unsuccessful, the origin of the event remains assigned to voxel i.

5. For every voxel i for which the $c_i > 0$ generate a sample from $p(d_i|c_i)$ using negative binomial distribution (Equation (6.44)). This step can be repeated many times to obtain several samples of $\mathbf{d}$ per one loop (steps 2 through 5) of the algorithm.

6. Repeat the last four steps. The last four steps constitute generation of a new Markov state in the Markov chain of states of $\mathbf{y}$ of the OE algorithm and at least one sample of $\mathbf{d}$ drawn from $p(\mathbf{d}|\mathbf{g})$.

We dene an iteration of this algorithm as steps 2 through 5 repeated N times (N is the total number of detected events).

The algorithm for sampling $p(\mathbf{r}|\mathbf{g})$ ($\mathbf{r}$ is the total number of nuclides in voxels) can be derived using identical steps as in the case of $p(\mathbf{d}|\mathbf{g})$ with the difference that instead of using $p(d_i|c_i)$ in step 5 the $p(r_i|c_i)$ is used. The

$$p(r_i|c_i) = \mathcal{N}B(r_i - c_i|c_i, 1 - \epsilon_i q) = \binom{r_i - 1}{r_i - c_i}(\epsilon_i q)^{c_i}(1 - \epsilon_i q)^{r_i - c_i} \quad (6.45)$$

where q is defined in Chapter 3 and is the chance that a nuclide decays during the experiment.

6.2 EXAMPLES OF STATISTICAL COMPUTING

In this section we provide examples of statistical computing based on Poisson-multinomial statistics. A very simple model of the tomographic nuclear imaging system is used that allows a clear presentation of the ideas developed in this book without additional complications from the description of imaging systems. These algorithms can be extended (albeit sometimes this extension may not be trivial) to real imaging systems and applications as described in Chapter 5.

6.2.1 SIMPLE TOMOGRAPHIC SYSTEM (STS)

The three-voxel model of tomographic system will be used to demonstrate various potential applications of the theory presented in this book. A similar system was already introduced in Chapter 3. Here, we slightly generalize the STS to K detector elements (when introduced in Example 3.2 K was fixed to 3). No restrictions on values of the elements of the system matrix are used. Using these assumptions the STS is visualized in Figure 6.5. The voxels are indicated by c_1, d_1, r_1 (voxel 1), c_2, d_2, r_2 (voxel 2), and c_3, d_3, r_3 (voxel 3) which correspond to the number of events that occurred in voxels and were detected (by either detector element) $\mathbf{c}$, the number of events that occurred (detected or not) $\mathbf{d}$, and the number of nuclides per voxel $\mathbf{r}$. There are K detector elements indicated by elements of the data vector $\mathbf{g}$. For each voxel all elements of the complete data vector $\mathbf{y}$ are also shown in Figure 6.5.

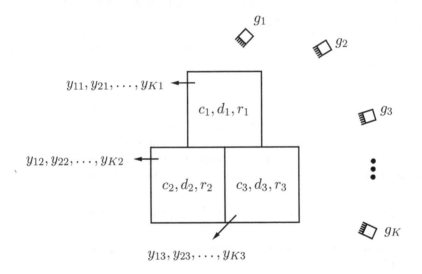

FIGURE 6.5 Schematics of the simple tomographic system (STS).

Using the definition of voxels we assume that if events occur in the same voxel they have the same chance of being detected in detector elements indexed by $k \in \{1, \ldots, K\}$. These chances are described by the system matrix

$$\hat{\alpha} = \begin{bmatrix} \alpha_{11} & \alpha_{12} & \alpha_{13} \\ \alpha_{21} & \alpha_{22} & \alpha_{23} \\ \cdots & \cdots & \cdots \\ \alpha_{K1} & \alpha_{K2} & \alpha_{K3} \end{bmatrix}. \tag{6.46}$$

The vector of complete data $\mathbf{y}$ has $3K$ elements that we denote as y_{ki}. As already emphasized in Chapter 3, although two indices are used to indicate the elements we insist that **y** is regarded as a vector.

> **Data generation:** In all examples we will only use the quantity **d** (number of events (decays) that occured) to demonstrate examples of the use of the methods developed in this book. The analysis of the other quantity of interest **r** which is the total number of radioactive nuclei in voxels at the begining of the scan follows identical steps as analysis of **d**. The only difference is that the elements of the system matrix need to be multiplied by q (defined in Section 3.5.1) which is a chance that the nuclide decays during the time of the experiment.
>
> One of the frequent tasks that will be performed in computer simulations is the data (**g**) generation. That is, if we assume that we know the state of nature in our case **d** we are to generate (simulate) **g** representing the measurement. As shown in Chapter 3, the data is a sample from Poisson-multinomial distribution. For our case of three-voxel imaging system this sample was generated by generating three independent samples (each of three samples was a $K + 1$ dimensional vector corresponding to three voxels) from multinomial distributions with $K + 1$ number of categories. As before, category 0 included events that were not detected and categories 1 through K corresponded to events detected in detectors 1 through K. For a given voxel i, the probabilities used in generating a multinomial sample were equal to the elements of the system matrix α_{ki} and for category 0, $1 - \sum_{k=1}^{K} \alpha_{ki} = 1 - \epsilon_i$. The samples from multinomial distribution were obtained using a numerical algorithm by Ahrens and Dieter [1].

6.2.2 IMAGE RECONSTRUCTION

Image reconstruction is a process of finding a point estimate of quantities of interest for each voxel representing the imaged volume. In our case we seek the estimate of **d** which is the number of events that occurred during the scan per voxel. If divided by imaging time this quantity may be considered as closely related to activity used in standard approaches [92] (it is not the same though). The image reconstruction using methods developed in this book provides point estimates with the errors. Instead of point estimate the estimator of interval of probable values can also be used. The example presented in this section details the methods required for obtaining such estimates. The error estimates of the image reconstruction or interval estimators are obtained with no significant additional computing cost.

We set up the following experiment. We assume that voxels 1, 2, and 3 emitted 2000, 2000, and 10000 events and therefore $\mathbf{d} = [2000, 2000, 10000]$. We assumed three detector elements $K = 3$ and the system matrix is

$$\begin{bmatrix} 0.1 & 0.1 & 0 \\ 0.1 & 0 & 0.1 \\ 0 & 0.1 & 0.1 \end{bmatrix}. \tag{6.47}$$

The sensitivity of each voxel is the same and is equal to 0.2. The sensitivity $\epsilon_i = \sum_{k=1}^{K} \alpha_{ki}$ is obtained by summing values of the system matrix in columns.

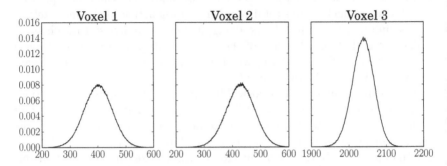

FIGURE 6.6   Marginalized distributions of $p(c_1|\mathbf{G} = \mathbf{g})$ (left), $p(c_2|\mathbf{G} = \mathbf{g})$ (middle), and $p(c_3|\mathbf{G} = \mathbf{g})$ (right).

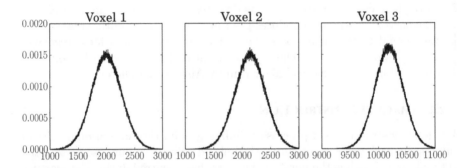

FIGURE 6.7   Marginalized distributions of $p(d_1|\mathbf{G} = \mathbf{g})$ (left), $p(d_2|\mathbf{G} = \mathbf{g})$ (middle), and $p(d_3|\mathbf{G} = \mathbf{g})$ (right).

We generated the data for the system and obtained $\mathbf{g} = [415, 1219, 1233]$. This vector constitutes the data and the task is to obtain an estimate of $\mathbf{d}$. We assumed a flat prior $p(\mathbf{d}) \propto 1$ and implemented OE algorithm given on page 192. We used 1000 iterations of of the OE algorithm as a burnin time, and 1,000,000 iterations to acquire 1,000,000 samples of the posterior. Samples of $\mathbf{c}$ and $\mathbf{d}$ were collected and saved.

The number of samples obtained for given values of c_1, c_2 and c_3 or d_1, d_2 and d_3 was divided by the total number of samples (1,000,000 in our case) to provide the estimate of the posterior probabilities $p(c_1, c_2, c_3|\mathbf{G} = \mathbf{g})$ and $p(d_1, d_2, d_3|\mathbf{G} = \mathbf{g})$. The estimates of marginalized posterior distributions $p(c_1|\mathbf{G} = \mathbf{g}), p(c_2|\mathbf{G} = \mathbf{g}), p(c_3|\mathbf{G} = \mathbf{g})$ are shown in Figure 6.6. The marginalized posterior distributions $p(d_1|\mathbf{G} = \mathbf{g}), p(d_2|\mathbf{G} = \mathbf{g}), p(d_3|\mathbf{G} = \mathbf{g})$ are shown

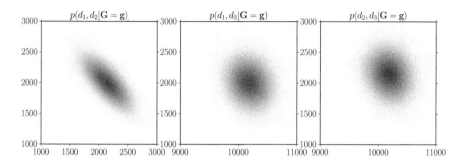

FIGURE 6.8 Marginalized distributions of $p(d_1, d_2 | \mathbf{G} = \mathbf{g})$ (left), $p(d_1, d_3 | \mathbf{G} = \mathbf{g})$ (middle), and $p(d_2, d_3 | \mathbf{G} = \mathbf{g})$ (right). Clearly, correlations between d_1 and d_2 (more oval shape of the distribution) are much stronger than correlations of quantity d_3 with either d_1 or d_2.

in Figure 6.7. To obtain the marginalized distributions from the estimate of the 3D posteriors is analogical to the marginalization of the probability wire cube—a concept introduced in Example 1.16 in Chapter 1. For the case considered here this is trivial. To obtain marginalized distribution of $p(c_1 | \mathbf{G} = \mathbf{g})$, the samples obtained by the OE algorithm are histogrammed considering only the value of c_1. Other two dimensions c_2 and c_3 are simply ignored. The 3D approximation of the posterior obtained by Monte Carlo methods can be marginalized not only to 1D distributions as shown in Figure 6.7, but also to 2D posterior distributions which can be used to illustrate the correlations between different quantities. In Figure 6.8 the three possible marginalizations of the estimate of the posterior are presented. In higher dimensional space the marginalizations can be performed to any lower dimensional space of any quantities of interest or functions of those quantities.

Image reconstruction provides just a single value of the estimate per voxel (point estimate). In order to find such estimate, the loss function need to be defined (see Chapter 2). For the example used here we assume quadratic loss function in which case the minimum mean square error (MMSE) estimate is obtained. The MMSE estimate is the mean of the posterior. The error of this estimate is calculated from the posterior variance defined and equal to the square root of the posterior variance.

The Bayesian analysis discussed in this book allows for range estimation as well which can be considered as a special case of image reconstruction. The range estimation was discussed in Section 2.3.2 and in our example we define it as 95% HDP (highest density posterior) confidence set. Table 6.1 lists all image reconstruction results for the example used here.

TABLE 6.1

Results of the image reconstruction for STS

	c_1	c_2	c_3
True	417	411	2039
MMSE[a]	401	429	2037
Error[b]	±50	±50	±29
95%CS[c]	(299, 494)	(299, 521)	(1976, 2090)

	d_1	d_2	d_3
True	2000	2000	10000
MMSE[a]	2006	2145	10185
Error[b]	±264	±263	±248
95%CS[c]	(1488, 2521)	(1420, 2653)	(9694, 10667)

[a] Minimum-Mean-Square Error estimator
[b] Square root of the variance of marginalized posterior
[c] CS=confidence set

6.2.3   BAYES FACTORS

Results shown in Figures 6.6 and 6.7 and Table 6.1 are obtained using image reconstruction methods with flat prior. Here we compare effects of other priors introduced in Section 6.1.2. We intentionally use priors which represent "inaccurate knowledge"[6] to demonstrate how it affects the analysis. The extension of these examples to accurate priors is straightforward.

Implementation of gamma prior: Implementation of the gamma prior requires knowledge about the mean of the number of events per voxel. Suppose that we have an educated guess about the number of emissions per voxel 1 and voxel 2 is 1900 and the number of emissions per voxel 2 is 11,000 (the true number was 2000 for voxel 1 and 2 and 10,000 for voxel 3). The confidence about these guesses is quantified by ω which we refer to as *gamma prior sensitivity*. Two confidence levels will be tested. The high confidence level denoted as 'hi' corresponds to $\omega = 1$ for voxels 1 and 2 and $\omega = 0.1$ for voxel 3. The low confidence level will be defined by ω equal to 0.02 for voxels 1 and 2 and 0.004 for voxel 3. Parameter ω of the gamma prior can be interpreted in some way as a value similar to sensitivity of the voxel. This is because in equations defining the ratio of posteriors obtained using gamma prior (Equations (6.34) and (6.35)) the value ω is directly added to voxel

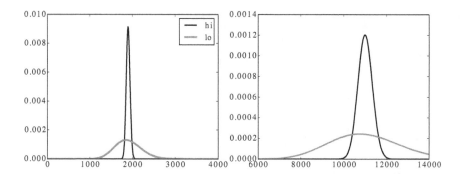

FIGURE 6.9 Gamma priors of **d** used in the experiment. On the left are gamma priors used for pixel values d_1 and d_2, and on the right the gamma priors used for values of pixel d_3. The 'hi' indicates high confidence in the prior (albeit incorrect) and 'lo' the low confidence in the prior.

sensitivities which is a reason for the name gamma prior sensitivity used here.

The priors are shown in Figure 6.9. We expect that using inaccurate prior with high confidence will typically lead to inaccurate posteriors as well and incorrect inferences. These expectations are consistent with the Bayesian philosophy which sees the experiment as a method of updating prior beliefs. If the experimental data summarized by the likelihood is strong enough, it may overcome the prior and provide posterior that described the true state of nature more accurately. Obviously, the data may take our correct (or incorrect) beliefs and "deceive us" if some "unusual" observation occurs changing the accurate prior beliefs to inaccurate posterior beliefs or making the inaccurate prior beliefs even more inaccurate. In other cases the experiential data may be weak and do not provide enough evidence to overcome the strength of prior beliefs and therefore the prior will not be changed significantly.

For the example used here of nuclear imaging with STS we started with $\mathbf{d} = [2000, 2000, 10000]$ and simulated the data acquisition which generated unobservable $\mathbf{c} = [417, 411, 2039]$ and observable $\mathbf{g} = [415, 1219, 1233]$. Let's examine how these observable data affects prior beliefs about $\mathbf{d}$ summarized in Figure 6.9. Both priors shown in the figure are inaccurate however one (denoted as 'hi') represents the high confidence and the other (denoted as 'lo') represents the low confidence in the beliefs. The high confidence is illustrated by the fact that the high-confidence prior is concentrated around few values that are believed to be true as opposed to the low-confidence prior for which a wider range of values is assumed probable.

---

[6]Inaccurate knowledge (prior) indicate that the true state of nature has low prior probability compared to other possibilities.

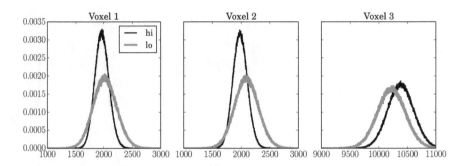

FIGURE 6.10 Marginalized posterior distributions of $p(d_1|\mathbf{G} = \mathbf{g})$ (left), $p(d_2|\mathbf{G} = \mathbf{g})$ (middle), and $p(d_3|\mathbf{G} = \mathbf{g})$ (right) for gamma priors. Posteriors marked with 'hi' and 'lo' indicate that the posterior was computed from initial high-confidence and low-confidence priors. The simulated true values of $\mathbf{d}$ were equal to 2000, 2000, 10000 for voxels 1, 2, 3, respectively.

The marginalized posteriors and shown in Figure 6.10. For the voxels 1 and 2 the posterior derived from the high-confidence prior is narrower than posteriors derived from the low-confidence prior which illustrates that the uncertainty in the prior beliefs still play important roles in posteriors. This is not so for voxel 3 for which more precise data was available (voxel 3 emitted 5 times more counts). The information in the prior was almost suppressed by information the data. The posteriors for voxel 3 are shifted and the value of the posterior for the true value (10,000) is quite small for the high confidence prior.

In order to offset the influence of inaccurate priors on the analysis we may use Bayes factors as an indicator of whether a given value of the number of events per voxel is supported by the data. The Bayes factors [8, 84], from definition, are equal to the ratio of the posterior of some quantity to the value of the prior of the same quantity. In our case the quantity is the number of emissions per voxel d.

In general the data in the example is quite strong and resulting posteriors are very similar regardless of quite different prior used to obtain them (Figure 6.9). We can conclude that analysis is robust. If the data is good prior does not matter. This point is investigated in the next section.

6.2.4   EVALUATION OF DATA QUALITY

In Bayesian analysis with real data (not computer simulation), recognizing if priors are accurate or inaccurate will be challenging since the true state of nature is unknown. If prior beliefs are confident (narrow prior) and inaccurate, the results of the analysis will also be inaccurate unless the strength of the data overcomes the strength of the prior. To some degree, this was illustrated in the previous section. The worse case scenario is to have confident and inaccurate priors. If possible, the prior beliefs should always be formulated broadly to prevent overconfidence (too narrow prior). It is better to have less precise analysis than to be wrong in most problems that involve uncertainty.

Aside from the obvious approach of obtaining more accurate priors, the universal method to counter the effect of inaccurate prior knowledge and to improve the Bayesian analysis is to obtain a better quality data. The better quality data is obtained by obtaining more counts or counts with better intricacy (Section 5.3.2). In nuclear imaging adding more counts does not necessarily result in significant improvement in the posteriors. Ideally each of the detector elements should register counts from as few voxels as possible or in other words as low intricacy as possible (see Section 5.3.2). The methods provided in this book allow studying various effects of sampling schemes on posteriors.

As an example we examine four different imaging systems by modifying the system matrix used in previous sections and given by Equation (6.47). The posteriors obtained by those systems are then compared. The four system matrices are:

1. $\hat{\alpha}$ as used in the previous section (Equation (6.47)).
2. $\hat{\alpha}_{IS}$ indicating increased sensitivity of all voxels. In the example sensitivity was twice as that used before.
3. $\hat{\alpha}_{AD}$ which was obtained by adding another detector element (row in the system matrix). The added detector acquired data from all three voxels with sensitivity of 0.2.
4. $\hat{\alpha}_{RE}$ is the system matrix for which all elements were divided by 2.

The explicit form of the system matrices is as follows

$$
\hat{\alpha} = \begin{bmatrix} 0.1 & 0.1 & 0 \\ 0.1 & 0 & 0.1 \\ 0 & 0.1 & 0.1 \end{bmatrix}, \quad \hat{\alpha}_{IS} = \begin{bmatrix} 0.2 & 0.2 & 0 \\ 0.2 & 0 & 0.2 \\ 0 & 0.2 & 0.2 \end{bmatrix},
$$

$$
\hat{\alpha}_{AD} = \begin{bmatrix} 0.1 & 0.1 & 0 \\ 0.1 & 0 & 0.1 \\ 0 & 0.1 & 0.1 \\ 0.2 & 0.2 & 0.2 \end{bmatrix}, \quad \hat{\alpha}_{RE} = \begin{bmatrix} 0.05 & 0.05 & 0 \\ 0.05 & 0 & 0.05 \\ 0 & 0.05 & 0.05 \end{bmatrix}. \tag{6.48}
$$

For all numerical experiments, as before, we used $\mathbf{d} = [2000, 2000, 10000]$. Posteriors were determined using 10^6 iterations of the OE algorithm. We used non-informative Jeffreys prior. For applications used in this work the flat prior and Jeffreys prior (acceptance ratios for Metropolis–Hastings algorithm are Equations (6.36) and (6.37)) provide an almost identical result. The effect of Jeffreys prior in nuclear imaging is more significant for a small (~ 10) number of events per voxel. It may play a more important role at a low count regime.

For the number of counts used in this example, which is in order to 1000, we have not detected any differences between the flat and Jeffreys prior.

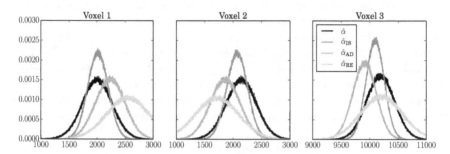

FIGURE 6.11 Marginalized posterior distributions of $p(d_1|\mathbf{G} = \mathbf{g})$ (left), $p(d_2|\mathbf{G} = \mathbf{g})$ (middle), and $p(d_3|\mathbf{G} = \mathbf{g})$ (right) for Jeffreys prior. Different posteriors correspond to different system matrices (different imagining systems) defined by Equation (6.48).

Figure 6.11 illustrated marginalized posteriors for different imaging systems. The effect of the increase ($\hat{\alpha}_{\text{IS}}$) or reduction ($\hat{\alpha}_{\text{RE}}$) of sensitivity compared to regular system matrix $\hat{\alpha}$ is quite obvious as the posteriors are narrower or wider indicating more and less informative results[7]. The effect of additional detector is much more subtle. It seems that it has only a major effect on voxel 3, but not as strong as the effect of the increase in the sensitivity for $\hat{\alpha}_{\text{IS}}$. This is an interesting finding because systems defined by $\hat{\alpha}_{\text{IS}}$ and $\hat{\alpha}_{\text{AD}}$ acquired the same number of counts, yet the benefit of these counts is different. This effect of weak improvement by adding the additional detector can be explained by worse intricacy of the AD imaging system compared to other systems and therefore worse performance for the same number of detected counts.

In Section 5.3.2 in Chapter 5 we defined intricacy of imaging system as a measure that can be used to classify the imaging system. Smaller values of intricacy was associated with a better imaging system. The intricacy ζ_k was defined for every detector element k (Equation (5.10)) as

$$\zeta_k = \sum_{k=1}^{K} \sum_{i=1}^{I} \frac{\alpha_{ki}}{\sum_{i'=1}^{I} \alpha_{ki'}} \log_2 \frac{\alpha_{ki}}{\sum_{i'=1}^{I} \alpha_{ki'}} \tag{6.49}$$

and the intricacy of imaging system was defined as the average of the intricacies of detector elements. Here we also define *posterior intricacy* $\zeta(\mathbf{g})$ which

---

[7]Because all posteriors have a bell-like shape to judge the width of the posterior, it is sufficient to examine the height of the posteriors which is quite straightforward. Posteriors with larger values at the peak are also narrower.

is defined as

$$\zeta_{\mathbf{g}} = \frac{1}{N} \sum_{k=1}^{K} \sum_{n=1}^{g_k} \zeta_k \tag{6.50}$$

which is the average intricacy per detected count. The system that achieves lower intricacy is considered better, as the average information entropy carried by detected counts is lower and therefore they provide more precise information about the location of the origin of the event. The intricacy is only one of the parameters of the system and does not fully reflect the imaging system capacity because the correlations between different counts (issue of tomographic sampling) are not accounted for in this measure. It is a useful and easy to compute measure, however, to compare the amount of information in the data.

The intricacy of $\hat{\alpha}_{\mathrm{IS}}$, $\hat{\alpha}_{\mathrm{IS}}$, $\hat{\alpha}_{\mathrm{RE}}$ is quite trivial to calculate because each detector element has the same ζ_k equal to $-\log_2 \frac{1}{2} = 1$. It follows that posterior intricacy is $\bar{\zeta} = \zeta_{\mathbf{g}} = 1$. For the $\hat{\alpha}_{\mathrm{AD}}$, the three detectors have intricacy of 1 and the additional detector has intricacy of $\log_2 3 = 1.58$. Therefore $\bar{\zeta} = 1.15$. The posterior intricacy specific to a given data set is equal to $\zeta_{\mathbf{g}} = 1.29$ which is an average between 1 and 1.58 as the half of the counts were detected in detectors with intricacy 1 and half of the counts in the detector with intricacy 1.58.

The increased intricacy of $\hat{\alpha}_{\mathrm{AD}}$ compared to $\hat{\alpha}_{\mathrm{IS}}$ is the reason for worse performance even though both systems have the same sensitivity.

6.2.5   DETECTION – BAYESIAN DECISION MAKING

One of the most powerful features of the Bayesian computing with the OE algorithm is that it allows to compute the posterior values of hypotheses and therefore it can use to a full extent the statistical information that is contained in the data and in the prior. For meaningful decision making, the loss function must be specified (Section 2.2) depending on the task that imaging system is designed to perform. The decision about the loss function must be taken on a case-by-case basis factoring in the quantified measures of consequences of the incorrect and correct decisions. In this section we demonstrate few examples of the application of the methods developed to provide a general approach towards decision problems. The methods can be extended to many other applications of the Bayesian decision theory in nuclear imaging.

Hot and cold lesion detections are one of the frequent tasks used in imaging. The hot/cold lesion is defined as increased/decreased concentration of a tracer in a voxel or in a group of voxels (region of interest, ROI). To illustrate this, let's use the STS and define the lesion as x-time increase in number of events in voxel 3 compared to the average concentration in the background. In our case the background is defined as the ROI comprising voxels 1 and 2. The hypothesis H_1 (lesion present in voxel 3) is that $d_3 > x \times \frac{1}{2}(d_1 + d_2)$ and hypothesis H_2 is that $d_3 \leq x \times \frac{1}{2}(d_1 + d_2)$. For clarity, we assume losses of incorrect decisions are equal to 1 and losses of correct decisions are 0. In such case we decide in favor of H_1 when $p(H_1|\mathbf{G} = \mathbf{g}) > p(H_2|\mathbf{G} = \mathbf{g})$ or equivalently $p(H_1|\mathbf{G} = \mathbf{g}) > 0.5$ (see Section 2.3.4 for more details). Table 6.2 presents the results of the lesion detection algorithm for imaging system defined by $\hat{\alpha}_{IS}$ and $\hat{\alpha}_{RE}$ shown in Equation (6.51). The $\hat{\alpha}_{IS}$ has twenty times more sensitivity than $\hat{\alpha}_{RE}$.

$$\hat{\alpha}_{IS} = \begin{bmatrix} 0.2 & 0.2 & 0 \\ 0.2 & 0 & 0.2 \\ 0 & 0.2 & 0.2 \end{bmatrix}, \quad \hat{\alpha}_{RE} = \begin{bmatrix} 0.01 & 0.01 & 0 \\ 0.01 & 0 & 0.01 \\ 0 & 0.01 & 0.01 \end{bmatrix}. \quad (6.51)$$

The true $\mathbf{d}$ was equal to $[2000, 2000, 10000]$ and therefore the true ratio between voxel 3 and background (average of voxels 1 and 2) was 5. The Flat, Jeffreys, high-confidence gamma (Section 6.2.3), and entropy with $\beta = 0.05$ priors were used.

TABLE 6.2

Posterior $p(H_1|\mathbf{G} = \mathbf{g})$ for various x[a]

| | Increased sensitivity (IS) | | | | |
| | | | x | | |
Prior	3	4	5	6	7
Flat	**1.00**	**1.00**	0.38	0.00	0.00
Gamma[b]	**1.00**	**1.00**	**0.88**	0.00	0.00
Jeffreys	**1.00**	**1.00**	0.39	0.00	0.00
Entropy[c]	**1.00**	**0.99**	0.00	0.00	0.00

| | Reduced sensitivity (RE) | | | | |
| | | | x | | |
Prior	3	4	5	6	7
Flat	**1.00**	**0.94**	**0.68**	0.34	0.12
Gamma[b]	**1.00**	**0.98**	**0.77**	0.37	0.11
Jeffreys	**1.00**	**0.96**	**0.73**	0.40	0.15
Entropy[c]	**0.99**	**0.83**	0.44	0.16	0.04

Note: Bold font indicates acceptance of hypothesis H_1.
[a] Hypothesis H_1 states that voxel 3 emitted x times more events than the average number of emitted events in the ROI defined by voxels 1 and 2. The loss function is assumed to be equal to 1 for incorrect decisions and 0 for correct decisions.
[b] High confidence inaccurate gamma prior defined in Section 6.2.3.
[c] $\beta = 0.95$.

The actual computation of the posterior of H_1 $p(H_1|\mathbf{g})$ is trivial once the posterior samples are determined by the OE algorithm. To compute the $p(H_1|\mathbf{g})$ we simply count the sampling points for which condition $d_3 > x \times \frac{1}{2}(d_1 + d_2)$ is true and divide by the number of sampling points. We count all the points neglecting the correlation between them because in the long run the correlations will average out if autocorrelation times are the same for different Markov chain states (Section 4.4.5). It was found that this indeed is the case in Sitek and Moore [96]. The Monte Carlo classical standard error of this estimate can be determined using bootstrap methods (Section 4.4.6).

Extension to multiple hypotheses: Based on the example of binary hypothesis testing the extension of the methods to multiple hypotheses is clear as long as whether the hypothesis is true or false can be ascertained for known $\mathbf{d}$. The $p(H_n|\mathbf{G} = \mathbf{g})$ is computed by counting the OE samples corresponding to a given hypothesis n and dividing by the total number of samples. Then, based on the values of the loss function, an appropriate decision (with the minimum expected loss) is selected as optimal.

Extension to unknown lesion locations: Based on this example, methods can be developed to compute the posterior of H_1 as a function of the hypothetical location of the lesions. Thus the approach can be used for finding the lesion location and lesion number as well. In this context it will be quite straightforward to implement the prior on the location and number of lesions. Such prior would reflect the belief that some regions in the image volumes can be classified as being more or less likely to contain lesions.

6.2.6 BAYES RISK

So far in this section, examples of the image reconstruction as well as the decision making process using posterior distributions were demonstrated. These were the examples of after-experiment decision making which were based solely on the posterior. We made assertions about the imaging systems based on just one data set that leads to the posterior. The criticism of such approach may be that findings for a given data may not apply for another data set and therefore cannot be generalized.

To remedy this, we introduced in Section 2.4.1 the Bayes risk which is the expected loss of a decision averaged over the marginalized distribution of data (Equation (2.32)). For certain decisions concerning repetitive use of some device on a class of objects, this is an excellent tool to grade data acquisition, image processing tools, and the decision rules. As explained in Section 2.4.1 the prior used when constructing the pre-posterior may not necessarily be the same as the distribution that defines the class of objects that are imaged. By π we indicated the prior used to form the posterior and by p we indicated distribution of objects. Following Equation (2.32) Bayes risk in terms of quantities used in this example is

$$\varpi(\delta^\pi) = \sum_{\mathbf{d}} p(\mathbf{d}) \sum_{\mathbf{g}} L(\mathbf{d}, \delta^\pi) p(\mathbf{g}|\mathbf{d}). \qquad (6.52)$$

The decision rule for which the Bayes risk is defined is δ^π. The sums $(\sum_{\mathbf{d}}, \sum_{\mathbf{g}})$ are performed over all vectors $\mathbf{d}$ and $\mathbf{g}$.

The easiest way to demonstrate how to use Bayes risk is to utilize the STS. We investigate two imaging systems 1 and 2 defined as

$$\hat{\alpha}_1 = \begin{bmatrix} 0.01 & 0.01 & 0 \\ 0.01 & 0 & 0.01 \\ 0 & 0.1 & 0.01 \end{bmatrix}, \quad \hat{\alpha}_2 = \begin{bmatrix} 0.005 & 0.005 & 0 \\ 0.005 & 0 & 0.005 \\ 0 & 0.005 & 0.005 \\ 0.01 & 0.01 & 0.01 \end{bmatrix}. \tag{6.53}$$

These systems are similar to IS and AD used in Section 6.2.4. They have the same sensitivity (the number of acquired counts is the same), but the intricacy of system 2 is higher which results in a worse performance than 1 as demonstrated in Section 6.2.4. The width of posteriors in Figure 6.11 were used to come to these conclusion. Here, we investigate those two systems (with reduced overall sensitivity) in terms of performance for the task of lesion detection as defined in the previous Section 6.2.5.

We define the lesion as voxel 3 having five times more events as the background comprising voxels 1 and 2. The loss function is the same as before and equal to 1 for correct decisions and to 0 for incorrect decisions. For this loss the optimal decision rule as shown in the previous section is to make a decision in favor of H_1 if $p(H_1|\mathbf{g}) > 0.5$. We will use this rule as δ^π in order to calculate the Bayes risk. In fact, we can use any other decision rule as well (e.g., decide lesion present for $p(H_1|\mathbf{g}) > 0.8$) to compute the Bayes risk, however, any other rule will give higher Bayes risk if $p(\mathbf{d}) = \pi(\mathbf{d})$ because it can be shown that Bayes risk is minimized for the decision rule that minimize expected loss. It is guaranteed that Bayes risk is minimized by the Bayes rule (which minimize expected loss) derived in the previous section. However, in the example analyzed here the $p(\mathbf{d}) \neq \pi(\mathbf{d})$ and therefore the Bayes rule is not guaranteed to minimize the Bayes risk.

We defined a class of objects being images in terms of the activity $\mathbf{f}$. This was done because it is convenient to specify gamma prior on $\mathbf{f}$ (see also Equation (6.54)).

$$p(\mathbf{f}) = \prod_{i=1}^{I} p(f_i) = \frac{1}{Z} \prod_{i=1}^{I} (f_i)^{\vartheta_i \omega_i - 1} e^{-\omega_i f_i}. \tag{6.54}$$

The $p(\mathbf{d})$ which describes the distribution of objects being imaged in terms of $\mathbf{d}$ can be easily derived from $p(\mathbf{f})$ as done in step 2 of the algorithm listed next. In the definition of $p(\mathbf{f})$ for voxels 1 and 2 we used $\vartheta_1 = \vartheta_2 = 2000$ and $\omega_1 = \omega_2 = 1$. For voxel 3 the $\vartheta_3 = 10000$ and $\omega_3 = 0.1$.

To compute the posterior and make a decision about rejecting/accepting the hypothesis we use the flat prior $\pi(\mathbf{d}) \propto 1$. With such defined imaging system and the task we compute the numerical estimate of Bayes risk ϖ for the imaging system 1 and imaging system 2 using the following steps:

1. Loop begins ($l = 0$).
2. Obtain a sample of the number of events per voxel **d**. This is done by first drawing a sample **f** from prior $p(\mathbf{f})$ and then obtaining a sample **d** drawn from the Poisson distribution $p(\mathbf{d}|\mathbf{f})$.
3. Obtain a sample of the data **g** from $p(\mathbf{g}|\mathbf{d})$ (Poisson-multinomial distribution).[8]
4. Compute samples of the posterior using OE algorithm. Use posterior obtained with $\pi(\mathbf{d})$ prior.
5. Compute $p(H_1|\mathbf{g})$ and decide in favor $(p(H_1|\mathbf{g}) > 0.5)$ or against $(p(H_1|\mathbf{g}) \leq 0.5)$ the hypothesis.
6. Compute the loss L_l where l is the index of the loop of the algorithm. If H_1 was decided but $d_3 \leq 5 \times \frac{1}{2}(d_1 + d_2)$ the loss is 1. If H_2 was decided but $d_3 > 5 \times \frac{1}{2}(d_1 + d_2)$ the loss is also 1. If decisions in favor of H_1 or H_2 were correct the loss is 0.
7. Loop ends ($l = l + 1$, repeat R times).
8. Compute the Bayes risk as $\varpi = \frac{1}{R}\sum_{l=1}^{R} L_l$.
9. (Optional) Compute classical standard error of ϖ using bootstrap.

Using the above algorithm, the Bayes risk was computed using 1000 samples of **d** ($R = 1000$), 1000 iterations of OE as a burning run, and 100,000 iterations to acquire samples of the posterior (1 sample per iteration). The 10,000 bootstrap samples were used to compute the estimate of the standard error of the Bayes risk. We found the $\varpi_1 = 0.122 \pm 0.011$ and $\varpi_2 = 0.262 \pm 0.014$, which confirms our previous assertions that system 1 is better (smaller Bayes risk) for the task of lesion detection.

---

[8]The result of steps 2 and 3 which is a sample from **g** can be obtained using other (simpler) algorithms as well.

A Probability distributions

We recall in this section common probability distributions used in this book. We also provide the means, variances, and skewness (for univariate distributions) of those distributions. The extensive review of probability distributions is provided by Johnson [50, 51] and Johnson et al. [52, 53].

A.1 UNIVARIATE DISTRIBUTIONS

A.1.1 BINOMIAL DISTRIBUTION

Binomial distribution describes the number of successes g in n Bernoulli trials with p being the chance of success

$$\mathcal{B}(g|n,p) = \binom{n}{g} p^g (1-p)^{n-g} \tag{A.1}$$

Mean=np, variance=$np(1-p)$, skewness=$\frac{1-2p}{\sqrt{np(1-p)}}$.

A.1.2 GAMMA DISTRIBUTION

Gamma distribution is important in applications of nuclear imaging because the prior expressed by the gamma distribution can be easily integrated in algorithms used in this book. Gamma distribution is continuous and has two parameters: $a > 0$ (shape parameters) and $b > 0$ (rate parameter):

$$\mathcal{G}(x|a,b) = \frac{b^a}{\Gamma(a)} x^{a-1} e^{-bx} \tag{A.2}$$

where $\Gamma(a)$ is the gamma function equal to $\int_0^\infty x^{a-1} e^{-x} dx$.

Mean=$\frac{a}{b}$, variance=$\frac{a}{b^2}$, skewness=$\frac{2}{\sqrt{a}}$.

A.1.3 NEGATIVE BINOMIAL DISTRIBUTION

Negative binomial distribution describes the number of successes g in Bernoulli trials if the number of failures $n > 0$ is known and probability of success is p

$$\mathcal{NB}(g|n,p) = \binom{g+n-1}{g} p^g (1-p)^n \tag{A.3}$$

Mean=$\frac{pn}{1-p}$, variance=$\frac{pn}{(1-p)^2}$, skewness=$\frac{1+p}{\sqrt{pn}}$

A.1.4 POISSON-BINOMIAL DISTRIBUTION

Poisson-binomial distribution describes the distribution of the sum of the n independent Bernoulli trials when trials are not identically distributed. Not identically distributed trials indicate that probabilities of success may be different for different trials and are denoted as $p_1, p_2, \ldots, p_n$. This distribution does not have a clean closed form of the distribution.

Mean=$\sum_{i=1}^{n} p_i$, variance=$\sum_{i=1}^{n}(1 - p_i)p_i$, skewness=(variance)$^{\frac{3}{2}} \sum_{i=1}^{n}(1 - 2p_i)(1 - p_i)p_i$.

A.1.5 POISSON DISTRIBUTION

Poisson distribution describes the probability of occurrence of a number of discrete events g during some time T if the rate of event occurrence is $\dot{f}(t)$. The rate $\dot{f}(t)$ is a deterministic function. We define f as

$$f = \int_0^T \dot{f}(t)dt \tag{A.4}$$

and the Poisson distribution is

$$\mathcal{P}(g|f) = \frac{f^g e^{-f}}{g!}. \tag{A.5}$$

The Poisson distribution can also be defined as a limit of the binomial distribution.
Mean=f, variance=f, skewness=$\frac{1}{\sqrt{f}}$.

A.1.6 UNIFORM DISTRIBUTION

$$\mathcal{U}(x|a, b) = \frac{1}{b - a}\chi_{[a,b]}(x) \tag{A.6}$$

where $\chi_{[a,b]}(x)$ is the indicator function equal to 1 for $x \in [a, b]$ and equal to 0 for $x \notin [a, b]$.

Mean=$\frac{1}{2}(a + b)$, variance=$\frac{1}{12}(b - a)^2$, skewness=0.

A.1.7 UNIVARIATE NORMAL DISTRIBUTION

Normal distribution is likely the most important probability distribution because of the central limit theorem discussed in Chapter 3.

$$\mathcal{N}(x|\mu, \sigma) = \frac{1}{\sqrt{2\pi}\,\sigma}e^{-\frac{(x-\mu)^2}{2\sigma^2}} \tag{A.7}$$

Mean=μ, variance=σ^2, skewness=0.

A.2 MULTIVARIATE DISTRIBUTIONS

In multivariate distributions the **g** is a vector of size K. The mean of the distribution is a vector of size K and covariance of the distribution is a symmetric positive-definite matrix of the size $K \times K$. Skewness is undefined for multivariate distributions.

A.2.1 MULTINOMIAL DISTRIBUTION

Multinomial distribution describes the number of successes in n categorical Bernoulli trials. The multinomial distribution is a generalization of the binomial distribution in which there are only two categories to which the Bernoulli trials can fall to. In categorical Bernoulli trials there are K categories and probability that the trial results in category k is p_k. A trial must provide the result and therefore $\sum_{k=1}^{K} p_k = 1$.

$$\mathcal{M}(\mathbf{g}|n, p_1, \dots, p_K) = \frac{n!}{g_1! \dots g_K!}(p_1)^{g_1} \dots (p_K)^{g_K} \tag{A.8}$$

$\text{Mean}(g_k) = np_k$, $\text{variance}(g_k) = np_k(1 - p_k)$, $\text{covariance}(g_k, g_{k'}) = -np_k p_{k'}$ for $k \neq k'$.

A.2.2 MULTIVARIATE NORMAL DISTRIBUTION

As in the case of the univarite normal distribution, the multivariate normal distribution is important because of the multivariate central limit theorem given in Chapter 3. If $\boldsymbol{\mu}$ is the vector of size K and $\boldsymbol{\Sigma}$ is the $K \times K$ positive-definite matrix then

$$\mathcal{N}(\mathbf{g}|\boldsymbol{\mu}, \boldsymbol{\Sigma}) = \frac{1}{\sqrt{(2\pi)^K |\boldsymbol{\Sigma}|^{-\frac{1}{2}}}} e^{-\frac{1}{2}(\mathbf{g}-\boldsymbol{\mu})^T \boldsymbol{\Sigma}^{-1}(\mathbf{g}-\boldsymbol{\mu})} \tag{A.9}$$

$\text{Mean}=\boldsymbol{\mu}$, $\text{variance}=\boldsymbol{\Sigma}$.

A.2.3 POISSON-MULTINOMIAL DISTRIBUTION

The Poisson-multinomial distribution (a.k.a. non-homogenous multinomial distribution) described the statistics of the sum of non-identically distributed categorical Bernoulli trials. We have K categories and n trials. By p_{ki} we denote the probability that in the ith trial the category k is chosen. These probabilities may not necessarily be equal for each trial. If they were equal, the Poisson-multinomial distribution would reduce to a simple multinomial distribution. A categorical trial must provide a result and therefore $\forall i \sum_{k=1}^{K} p_{ki} = 1$.

The Poisson-multinomial distribution does not have a closed form of the probability distribution.

$\text{Mean}(g_k) = \sum_{i=1}^{n} p_{ki}$, $\text{variance}(g_k) = \sum_{i=1}^{n} p_{ki}(1 - p_{ki})$, $\text{covariance}(g_k, g_{k'}) = -\sum_{i=1}^{n} p_{ki} p_{k'i}$ for $k \neq k'$.

B Elements of set theory

A *set* is a collection of distinct objects (elements) (for example, numbers, letters, vectors, other sets, etc.). The number of objects in the set (*cardinality* of the set) can either be finite and infinite. The sets are denoted by capital letters (e.g., $W, F, G, \mathbf{G}$). We use a notational convention and indicate by bold capital letters sets that are vectors. The set of vectors $\mathbf{G}$ can be considered as a set that contains elements that are also sets. The set can be defined by a description. For example, we define F as a set of all real numbers from 0 to 1, or G is the set of all non-negative integers. A set can also be defined explicitly by using curly brackets. For example,

$$W = \{1, 2, 3, 4\} \tag{B.1}$$

has four elements which are integer numbers 1 through 4. Another example could be

$$W = \{\text{'Yes'}, \text{'No'}\} \tag{B.2}$$

which has two elements labeled as 'Yes' and 'No'.

If a set has many elements the definition can be abbreviated by using ellipsis that is indicative of a continuity and the set elements can be deduced in some obvious way. For example

$$W = \{1, 2, \ldots, 100\} \tag{B.3}$$

is the definition of a hundred-element set with elements that are integer numbers from 1 to 100. Another way to specify the elements of the set is to provide descriptive specification inside the curly bracket as

$$W = \{n^2 : n \text{ is an integer}; 1 < n < 5\} \tag{B.4}$$

which is an equivalent to a three-element set $\{4, 9, 16\}$. In Equation (B.4) the colon can be read as "such that" and semicolon separates the conditions which elements of the set must satisfy.

If elements of one set are identical to elements of another set, the sets are equal which is indicated by

$$\{n^2 : n \text{ is an integer}; 1 < n < 5\} = \{4, 9, 16\} \tag{B.5}$$

Suppose an element from set W is denoted as ν. The membership of ν in W is indicated by $\nu \in W$, and if an object ν is not an element of W it is indicated by $\nu \notin W$. The *empty set* W is a set that does not contain any element and it is indicated by $W = \emptyset$.

A set W' is the *subset* of W when for any element $w \in W'$ the w is also an element of W. The relationships between subsets can be visualized using *Venn diagrams* shown in Figure B.1. The diagrams provide intuitive representation of sets inter-relations.

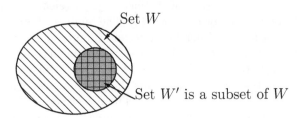

FIGURE B.1 The set W' is a subset of the set W. Alternatively we can say that W contains W'.

The notation $W' \subset W$ or $W \supset W'$ indicates that W' is a *proper subset* of W and that W is a *proper superset* of W', respectively. Both indicate that all elements of W' are also elements of W, but at least one element in W is not in W'. If the last condition does not have to be met we say that W' is a *subset* of W and that W is a *superset* of W' and indicate this by $W' \subseteq W$ or $W \supseteq W'$.

The sets can be "added" together. In the set theory this is called the *union*. If an element is in both sets that are added only a single "copy" of the elements is an element of the union. Some basic properties of the union are

$$W \cup W' = W' \cup W \tag{B.6}$$
$$W \cup (W' \cup W) = (W \cup W') \cup W'' \tag{B.7}$$
$$W \subseteq (W \cup W') \tag{B.8}$$

The *intersection* of two sets is defined as a set of elements that are elements of both sets that are being intersected. The intersection has the following properties (for a thorough coverage of theory of sets refer to Jech [48]).

$$W \cap W' = W' \cap W \tag{B.9}$$
$$W \cap (W' \cap W) = (W \cap W') \cap W'' \tag{B.10}$$
$$W \supseteq (W \cup W') \tag{B.11}$$

Sets W and W' are *disjoint* when $W \cap W' = \emptyset$. The set W can be *partitioned* into a collection of sets[1] $W_1, W_2, \ldots W_N$ if and only if

$$W = W_1 \cup W_2 \cup \ldots \cup W_N \tag{B.12}$$
$$W_i \cup W_j = \emptyset \text{ for } i \neq j \tag{B.13}$$

---

[1] The partition collection of sets is also a set.

C Multinomial distribution of single-voxel imaging

In this appendix we derive the distribution $p(\mathbf{y}_1|d_1)$ from the basic principles where $\mathbf{y}_1$ is the vector of the number of detected events originated in a voxel 1 and d_1 is the number of decays in voxel 1. Therefore it is a model of a simple imaging system with many detector elements and a single voxel. A figure showing a schematic of such a system is shown in Chapter 3 in Figure 3.5.

The derivation may seem at first quite complex and convoluted, and the result quite obvious. We strongly encourage, however, careful study of the appendix, as it provides insights into a general approach on handling the statistics of the photon limited data.

Using the chain rule of conditional probabilities, the $p(\mathbf{y}_1|d_1)$ can be specified as follows

$$
\begin{aligned}
p(\mathbf{y}_1|d_1) &= p\Big((\mathbf{y}_1)_1, (\mathbf{y}_1)_2, \ldots, (\mathbf{y}_1)_K \Big| d_1\Big) \\
&= p\Big((\mathbf{y}_1)_1 \Big| (\mathbf{y}_1)_2, \ldots (\mathbf{y}_1)_K, d_1\Big) p\Big((\mathbf{y}_1)_2 \Big| (\mathbf{y}_1)_3, \ldots (\mathbf{y}_1)_K, d_1\Big) \\
&\quad \times p\Big((\mathbf{y}_1)_3 \Big| (\mathbf{y}_1)_4 \ldots, (\mathbf{y}_1)_K, d_1\Big) \ldots p\Big((\mathbf{y}_1)_K \Big| d_1\Big). \quad \text{(C.1)}
\end{aligned}
$$

In the above equation we have the product of conditionals of a general form $p\Big((\mathbf{y}_1)_k | (\mathbf{y}_1)_{k+1} \ldots (\mathbf{y}_1)_K, d_1\Big)$. Considering one of such conditionals

$$
p\Big((\mathbf{y}_1)_k | (\mathbf{y}_1)_{k+1} \ldots (\mathbf{y}_1)_K, d_1\Big) \quad \text{(C.2)}
$$

the number of detections in detector elements $k+1, \ldots, K$ and the number of decays d_1 are assumed known. Therefore, based on the previous result (Equation (3.14)) where it was shown that the number of detections is binominally distributed given the number of decays we conclude that the above is binominally distributed as well. However, since the number of decays that is detected in detectors $k+1$ through K is known in the conditional, there is only a $d_1 - \sum_{k'=k+1}^{K}(\mathbf{y}_1)_{k'}$ decays that can be detected in detector k. Also, because detections at detectors $1 \ldots k$ are only considered (numbers of detections in other detectors are assumed known) the chance of detection in detector k has to be adjusted so the chances of detections in detectors $1, \ldots, k$ added to the chance of the decay being undetected adds to 1. This can be done by adjusting α_{k1} to α'_{k1} using the following:

$$
\alpha'_{k1} = \frac{\alpha_{k1}}{\Big(1 - \sum_{k'=k+1}^{K} \alpha_{k'1}\Big)} \quad \text{(C.3)}
$$

where α_{k1} indicates the chance that if the decay occurs in a voxel (indexed with 1) it will be detected in detector element k. The α'_{k1} is adjusted chance that takes into account that no detections are considered in detector elements $k+1, \ldots, K$.

Using the above $p(\mathbf{y}_1|d_1)$ (Equation (C.1)) becomes

$$p(\mathbf{y}_1|d_1) = \prod_{k=1}^{K} \binom{d_1 - \sum_{k'=k+1}^{K}(\mathbf{y}_1)_{k'}}{(\mathbf{y}_1)_k} \times$$

$$\times \left(\frac{\alpha_{k1}}{1 - \sum_{k'=k+1}^{K}\alpha_{k'1}}\right)^{(\mathbf{y}_1)_k} \left(1 - \frac{\alpha_{k1}}{1 - \sum_{k'=k+1}^{K}\alpha_{k'1}}\right)^{d_1 - \sum_{k'=k+1}^{K}(\mathbf{y}_1)_{k'} - (\mathbf{y}_1)_k} .$$

Simplifying the last bracket and using $\sum_{k'=k+1}^{K}(\mathbf{y}_1)_{k'} + (\mathbf{y}_1)_k = \sum_{k'=k}^{K}(\mathbf{y}_1)_{k'}$ it transforms to

$$p(\mathbf{y}_1|d_1) = \prod_{k=1}^{K} \binom{d_1 - \sum_{k'=k+1}^{K}(\mathbf{y}_1)_{k'}}{(\mathbf{y}_1)_k} \times$$

$$\times \left(\frac{\alpha_{k1}}{1 - \sum_{k'=k+1}^{K}\alpha_{k'1}}\right)^{(\mathbf{y}_1)_k} \left(\frac{1 - \sum_{k'=k}^{K}\alpha_{k'1}}{1 - \sum_{k'=k+1}^{K}\alpha_{k'1}}\right)^{d_1 - \sum_{k'=k}^{K}(\mathbf{y}_1)_{k'}} . \tag{C.4}$$

Although the above appears to be a complex expression, it can be considerably simplified using the following substitution

$$x_k \triangleq \frac{1}{\left(d_1 - \sum_{k'=k}^{K}(\mathbf{y}_1)_{k'}\right)!} \left(1 - \sum_{k'=k}^{K}\alpha_{k'1}\right)^{d_1 - \sum_{k'=k}^{K}(\mathbf{y}_1)_{k'}} . \tag{C.5}$$

With x_k Equation (C.4) becomes

$$p(\mathbf{y}_1|d_1) = \prod_{k=1}^{K} \frac{(\alpha_{k1})^{(\mathbf{y}_1)_k}}{(\mathbf{y}_1)_k!} \frac{x_k}{x_{k+1}} \tag{C.6}$$

All x_k's are canceled in the above with the exception of x_1 in the numerator and x_{k+1} in the denominator which are from the definition of x_k in Equation (C.5) equal to

$$x_1 = \frac{1}{\left(d_1 - \sum_{k'=1}^{K}(\mathbf{y}_1)_{k'}\right)!}(1 - \epsilon_1)^{d_1 - \sum_{k'=1}^{K}(\mathbf{y}_1)_{k'}} \tag{C.7}$$

and

$$x_{k+1} = \frac{1}{d_1!} \tag{C.8}$$

Defining also the voxel sensitivity ϵ_1 as $\sum_{k=1}^{K} \alpha_{k1}$, Equation (C.6) is transformed to the final distribution given by

$$p(\mathbf{y}_1|d_1) = \frac{d_1!}{\left(d_1 - \sum_{k'=1}^{K}(\mathbf{y}_1)_{k'}\right)! \prod_{k'=1}^{K}(\mathbf{y}_1)_{k'}!}$$

$$\times (1 - \epsilon_1)^{d_1 - \sum_{k'=1}^{K}(\mathbf{y}_1)_{k'}} \prod_{k=1}^{K} (\alpha_{k1})^{(\mathbf{y}_1)_k} \quad (C.9)$$

The above equation describes a *multinomial distribution* (something we suspected from the start) with $K+1$ categories which include K categories corresponding to detection of decays from voxel 1 in K number of detection elements and one category corresponding to decays that were not detected. The term $\frac{d_1!}{(d_1 - \sum_{k'=1}^{K}(\mathbf{y}_1)_{k'})! \prod_{k'=1}^{K}(\mathbf{y}_1)_{k'}!}$ corresponds to a number of ways d_1 decays can be distributed among $K+1$ categories with $d_1 - \sum_{k'=1}^{K}(\mathbf{y}_1)_{k'}, (\mathbf{y}_1)_1, \ldots (\mathbf{y}_1)_K$ counts in each of $K+1$ number of categories. The term $1 - \epsilon_1$ is a chance that a decay is not detected by the camera.

D Derivations of sampling distribution ratios

Derivation of sampling ratios stated in Equations (6.23)–(6.26) is provided here. We consider here only the flat sampling distribution corresponding to the flat prior $p_F(s)$ defined in Equation (6.6). The ratios of other distributions can be calculate similarly. In the flat prior we replace terms $c_i!$ with $\Gamma(c_i + 1)$ to have closer correspondence with the other priors. This comes from the property of the gamma function such $\Gamma(x+1) = x!$ for any non-negative integer x. We begin by stating the ratio of sampling distributions for state u and s using Equation (6.8). The definition of those states is given in Section 6.1.1. The states differ by a location of origin of a single event that was detected in detector element k and in the state s is located in voxel i and in state u it is located in voxel i'. We assume that α_{ki} and $\alpha_{ki'}$ are both greater than 0.

$$
\frac{p(u)}{p(s)} = \left[\frac{\frac{1}{Z} \prod_{j=1, j \neq i, j \neq i'} \frac{\Gamma(c_j+1)}{(\epsilon_j)^{c_j}} \prod_{l=1}^{K} \frac{(\alpha_{lj})^{y_{lj}}}{y_{lj}!}}{\frac{1}{Z} \prod_{j=1, j \neq i, j \neq i'} \frac{\Gamma(c_j+1)}{(\epsilon_j)^{c_j}} \prod_{l=1}^{K} \frac{(\alpha_{lj})^{y_{lj}}}{y_{lj}!}} \right]
$$
$$
\times \frac{\left[\frac{\Gamma(c_i)}{(\epsilon_i)^{c_i-1}} \frac{(\alpha_{ki})^{y_{ki}-1}}{(y_{ki}-1)!} \prod_{l=1, l \neq k}^{K} \frac{(\alpha_{li})^{y_{li}}}{y_{li}!} \right] \left[\frac{\Gamma(c_{i'}+2)}{(\epsilon_{i'})^{c_{i'}+1}} \frac{(\alpha_{ki'})^{y_{ki'}+1}}{(y_{ki'}+1)!} \prod_{l=1, l \neq k}^{K} \frac{(\alpha_{li'})^{y_{li'}}}{y_{li'}!} \right]}{\left[\frac{\Gamma(c_i+1)}{(\epsilon_i)^{c_i-1}} \frac{(\alpha_{ki})^{y_{ki}}}{y_{ki}!} \prod_{l=1, l \neq k}^{K} \frac{(\alpha_{li})^{y_{li}}}{y_{li}!} \right] \left[\frac{\Gamma(c_{i'}+1)}{(\epsilon_{i'})^{c_{i'}-1}} \frac{(\alpha_{ki'})^{y_{ki'}}}{y_{ki'}!} \prod_{l=1, l \neq k}^{K} \frac{(\alpha_{li'})^{y_{li'}}}{y_{li'}!} \right]}
$$
$$
\tag{D.1}
$$

The first square bracket in the equation above contains identical numerator and denominator and will be canceled. Expressions under the product symbol $\prod$ in numerator and denominator of the second part of the equation are identical and will be canceled as well. Using the property of the gamma function $\Gamma(x+1) = x\Gamma(x)$ for any real value of x and simplifying the remaining terms, we arrive at Equation (6.23) repeated here for convenience:

$$
\frac{p(u)}{p(s)} = \frac{p_F(u)}{p_F(s)} = \frac{\alpha_{ki'} \epsilon_i (c_{i'} + 1) y_{ki}}{\alpha_{ki} \epsilon_{i'} c_i (y_{ki'} + 1)}
\tag{D.2}
$$

E Equation (6.11)

Considering voxel i for which the expected number of events is f_i, the actual number of events in an experiment is distributed according to Poisson distribution

$$p(d_i|f_i) = \frac{(f_i)^{d_i} e^{-f_i}}{d_i!} \tag{E.1}$$

When event occurs it has the ϵ_i chance of being detected by the camera and therefore the total number of detections given the number of trials d_i $p(c_i|d_i)$ has binomial distribution as

$$p(c_i|d_i) = \binom{d_i}{c_i} (\epsilon_i)^{c_i} (1 - \epsilon_i)^{d_i - c_i} \tag{E.2}$$

It follows that

$$p(c_i|f_i) = \sum_{d_i=0}^{\infty} p(c_i|d_i, f_i) p(d_i|f_i) = \sum_{d_i=0}^{\infty} p(c_i|d_i) p(d_i|f_i) \tag{E.3}$$

where we used statistical independence of c_i and f_i given d_i. Since $p(c_i|d_i) = 0$ for $c_i > d_i$ using Equations (E.1) and (E.2) the above transforms to

$$p(c_i|f_i) = \sum_{d_i=c_i}^{\infty} \frac{(f_i)^{d_i} e^{-f_i}}{d_i!} \frac{d_i!}{(d_i - c_i)! c_i!} (\epsilon_i)^{c_i} (1 - \epsilon_i)^{d_i - c_i} \tag{E.4}$$

Using a substitution, $d_i' = d_i - c_i$, we obtain

$$p(c_i|f_i) = \frac{(f_i \epsilon_i)^{c_i} e^{-f_i}}{c_i!} \sum_{d_i'=0}^{\infty} \frac{(f_i - f_i \epsilon_i)^{d_i'}}{d_i'!} = \frac{(f_i \epsilon_i)^{c_i} e^{-f_i}}{c_i!} e^{f_i - f_i \epsilon_i} = \frac{(f_i \epsilon_i)^{c_i} e^{-f_i \epsilon_i}}{c_i!} \tag{E.5}$$

F C++ OE code for STS

```
class STS
{
public:
  // voxel number (I)
  static const int VOX = 3;
  // detector elements (K)
  static const int DE = 3;
  enum PRIORTYPE {FLAT=0, GAMMA, JEFFREYS, ENTROPY};
  // These are statistical functions used / code is not provided
  ArekStat *stat;
private:

  PRIORTYPE  prior;
  // prior parameters for gamma prior
  double g1[VOX],g2[VOX];
  // prior parameter for entropy prior
  double beta;

  // lookups
  // Size if equal to number of counts

  // detector bin in which count was detected
  int *lookupBins;
  // voxel in which the origin is located
  int *lookupVoxels;

  // data
  int g[DE];

  //  Number of events emitted and detected per voxel
  int c[VOX];
  //         Number of event emitted
  int d[VOX];
  // Number of radioactive nuclei in voxel
  double decay_chance;
  int r[VOX];

  // system matrix
  double sm[VOX*DE];
#define SM(k,i) sm[k*VOX+i]
  // system matrix normalized for each detector element
  //it has cumulative values in rows. The last value in row must be 1
  double smN[VOX*DE];
#define SMN(k,i) smN[k*VOX+i]
  // sensitivity
  double sens[VOX];

private:
  void InitObject()
  {
    for(int i=0;i<VOX;i++)
    {
      sens[i]=0;
      for(int k=0;k<DE;k++)
      {
        sens[i] += SM(k,i);
      }
    }

    memcpy(smN,sm,sizeof(double)*VOX*DE);
```

```
  for(int k=0;k<DE;k++)
  {
    double norm = 0;
    for(int i=0;i<VOX;i++)
      norm += SM(k,i);
    SMN(k,0) = SM(k,0)/norm;
    for(int i=1;i<VOX;i++)
      SMN(k,i) = SMN(k,i-1)+SM(k,i)/norm;
  }

  memset(g,0,DE*sizeof(int));
  memset(c,0,VOX*sizeof(int));
  memset(d,0,VOX*sizeof(int));

  stat = new ArekStat;
  lookupBins=0;
  lookupVoxels=0;
  prior = FLAT;

  // If gamma prior is used the specs are here
  g1[0] = g1[1] = 1900;
  g1[2] = 11000;
  g2[0] = g2[1] = 1;
  g2[2] = 0.1;

  // If entropy prior is used the specs are here
  beta = 0.95;
  // The r (total number of radioactive nuclei
  // is of interest this value has to be set to chance
  // that nuclei decays during imaging time.
  decay_chance = 1.;
}

public:
  // Constructor
  STS(double *sminit)
  {
    memcpy(sm, sminit, sizeof(double)*VOX*DE);
    InitObject();
  };
  // called when data is initialized
  void InitRecon()
  {
    for(int i=0;i<VOX;i++) c[i]=0;
    int count = 0;
    for(int bin=0;bin<DE;bin++)
    {
      for(int n=0;n<g[bin];n++)
      {
        int pixel = Select(bin);
        c[pixel]++;
        lookupBins[count] = bin;
        lookupVoxels[count++] = pixel;
      }
    }
    Compute_d();
    if(decay_chance < 1.)
    {
      Compute_r();
    }
    else
    {
      memset(r,0,sizeof(int)*VOX);
    }
  }
  // initialize the data
```

```
void Set_g(int *gg)
{
  for(int k=0;k<DE;k++) g[k]=gg[k];
  int sum=0;
  for(int k=0;k<DE;k++)  sum+=g[k];
  if(lookupBins) delete [] lookupBins;
  lookupBins = new int[sum];
  if(lookupVoxels) delete [] lookupVoxels;
  lookupVoxels = new int[sum];
  InitRecon();
}

// Generates the data g from c  (forward model)
void Make_g_from_c(int *cc)
{
  memset(g,0,sizeof(int)*DE);

  for(int i=0;i<VOX;i++)
  {
    double ps[DE];
    // normalize to 1
    for(int k=0;k<DE;k++)
    {
      ps[k] = SM(k,i) / sens[i];
    }
    // multinomial sample
    int *y = stat->i4vec_multinomial_sample(cc[i],ps,DE);
    for(int k=0;k<DE;k++)
    {
      g[k] += y[k];
    }
    delete [] y;
  }
  int sum = 0;
  for(int k=0;k<DE;k++)
  {
    sum += g[k];
  }
  if(lookupBins) delete [] lookupBins;
  lookupBins = new int[sum];
  if(lookupVoxels) delete [] lookupVoxels;
  lookupVoxels = new int[sum];
  InitRecon();
}
// Generate data from d (forward model)
void Make_g_from_d(int *dd)
{
  memset(g,0,sizeof(int)*DE);
  for(int i=0;i<VOX;i++)
  {
    double ps[DE+1];
    for(int k=0;k<DE;k++)
    {
      ps[k] = SM(k,i);
    }
    ps[DE]=1-sens[i];
    // multinomial sample
    int *y = stat->i4vec_multinomial_sample(dd[i],ps,DE+1);
    for(int k=0;k<DE;k++)
    {
      g[k] += y[k];
    }
    delete [] y;
  }
  int sum = 0;
  for(int k=0;k<DE;k++)
```

```
  {
    sum += g[k];
  }
  if(lookupBins) delete [] lookupBins;
  lookupBins = new int[sum];
  if(lookupVoxels) delete [] lookupVoxels;
  lookupVoxels = new int[sum];
  InitRecon();
}

// Generate data from r (forward model)
void Make_g_from_r(int *rr, double ini_decay_chance)
{
  memset(g,0,sizeof(int)*DE);
  decay_chance = ini_decay_chance;
  for(int i=0;i<VOX;i++)
  {
    double ps[DE+1];
    for(int k=0;k<DE;k++)
    {
      ps[k] = SM(k,i) * decay_chance;
    }
    ps[DE]=1-sens[i] * decay_chance;
    int *y = stat->i4vec_multinomial_sample(rr[i],ps,DE+1);
    for(int k=0;k<DE;k++)
    {
      g[k] += y[k];
    }
    delete [] y;
  }
  int sum = 0;
  for(int k=0;k<DE;k++)
  {
    sum += g[k];
  }
  if(lookupBins) delete [] lookupBins;
  lookupBins = new int[sum];
  if(lookupVoxels) delete [] lookupVoxels;
  lookupVoxels = new int[sum];

  InitRecon();
}
// Select step of MCMC
// for a counts detected in detector bin
// find random voxel with chance proportional to
// the value of system matrix element
int Select(int bin)
{
  double *p = smN + bin*VOX;
  double rnd = stat->r8_uniform_sample(0,1);
  for(int i=0;i<VOX;i++)
  {
    if(rnd<=p[i])
    {
      return i;
    }
  }
  // this should never happen
  return -1;
}

// Accept step of MCMC
// to = voxel where event is being moved to
// from = voxel where event is being moved from
bool Accept(int to, int from)
{
```

```
  double ratio=0;
  if(prior==FLAT)
  {
    ratio = (double)(c[to]+1)/sens[to]/(double)c[from]*sens[from];
    if(ratio>=1) return true;
  }
  else if(prior==GAMMA)
  {
    ratio = (double)(c[to]+g1[to]*g2[to])/(sens[to]+g2[to])/
                          (double)(c[from]+g1[from]*g2[from]-1)*(sens[from]+g2[from]);
  }
  else if(prior==JEFFREYS)
  {
    ratio = ((double)(c[to])+0.5)/sens[to]/((double)c[from]-0.5)*sens[from];
    if(ratio>=1) return true;
  }
  else if(prior==ENTROPY)
  {
    ratio = ((double)(c[to])+1.0)/sens[to]/((double)c[from])*sens[from];
    if(ratio>=1) return true;
    ratio = pow(ratio,beta);
  }

  if(stat->r8_uniform_01_sample() < ratio)
  {
    return true;
  }
  return false;
}

// Get sample of d given c
void Compute_d()
{
  for(int i=0;i<VOX;i++)
  {
    d[i] = stat->neg_binom(c[i],sens[i]);
  }
}
// Get sample of r given r
void Compute_r()
{
  for(int i=0;i<VOX;i++)
  {
    r[i] = stat->neg_binom(c[i],sens[i]*decay_chance);
  }
}


// Main function. Runs 'iter' number of iterations of the OE algorithm
void Run(int iter)
{
  int tot = 0;
  for(int k=0;k<DE;k++)
  {
    tot += g[k];
  }
  for(int it=0;it<iter;it++)
  {
    for(int ii=0;ii<tot;ii++)
    {
      // get one of the counts.
      int count = (int)( stat->r8_uniform_01_sample()*(double)tot );
      // location of origin for current Markov state
      int old_pixel = lookupVoxels[count];
      // detector bin where count was detected
      int bin = lookupBins[count];
```

```
      // randomly determine a candidate voxel for the origin
      int new_pixel = Select(bin);
      // if by chance it is the same voxel just skip
      if(new_pixel==old_pixel) continue;
      // If accepted ...
      if(Accept(new_pixel, old_pixel))
      {
        // change location of the origin of the count
        lookupVoxels[count] = new_pixel;
        // update
        c[new_pixel]++;
        c[old_pixel]-;
      }

    }
  }
 }
};
```

References

1. J. H. Ahrens and U. Dieter. Computer methods for sampling from gamma, beta, poisson and bionomial distributions. *Computing*, 12(3): 223–246, September 1974.
2. H. O. Anger. Scintillation camera. *Review of Scientific Instruments*, 29 (1):27–33, January 1958.
3. H. O. Anger. Scintillation camera with multichannel collimators. *Journal of Nuclear Medicine*, 5(7):515–531, July 1964.
4. P. J. Aston. Is radioactive decay really exponential? *EPL (Europhysics Letters)*, 97(5):52001, March 2012. arXiv: 1204.5953.
5. D. L. Bailey, David W. Townsend, Peter E. Valk, and Michael N. Maisey. *Positron Emission Tomography: Basic Sciences*. Springer, New York, 2003 edition, July 2005.
6. H. H. Barrett and K. J. Myers. *Foundations of Image Science*. Wiley-Interscience, Hoboken, NJ, 1st edition, October 2003.
7. F. J. Beekman, F. van der Have, B. Vastenhouw, A. J. A. van der Linden, P. P. van Rijk, J. P. H. Burbach, and M. P. Smidt. U-SPECT-i: A novel system for submillimeter-resolution tomography with radiolabeled molecules in mice. *Journal of Nuclear Medicine*, 46(7):1194–1200, July 2005.
8. J. O. Berger. *Statistical Decision Theory and Bayesian Analysis*. Springer, New York, December 2010.
9. M. Berman, S. M. Grundy, and B. V. Howard, editors. *Lipoprotein Kinetics and Modeling*. Academic Press, January 1982.
10. J. M. Bernardo. Reference posterior distributions for bayesian inference. *Journal of the Royal Statistical Society. Series B (Methodological)*, 41 (2):113–147, January 1979.
11. G. E. P. Box. Sampling and bayes' inference in scientific modelling and robustness. *Journal of the Royal Statistical Society. Series A (General)*, 143(4):383, 1980.
12. G. E. P. Box and M. E. Muller. A note on the generation of random normal deviates. *The Annals of Mathematical Statistics*, 29(2):610–611, June 1958.
13. George E. P. Box and G. C. Tiao. *Bayesian Inference in Statistical Analysis*. Wiley-Interscience, New York, 1st edition, April 1992.
14. S. P. Brooks and A. Gelman. General methods for monitoring convergence of iterative simulations. *Journal of Computational and Graphical Statistics*, 7(4):434–455, December 1998.
15. R. R. Buechel, B. A. Herzog, L. Husmann, I. A. Burger, A. P. Pazhenkottil, V. Treyer, I. Valenta, P. von Schulthess, R. Nkoulou, C. A. Wyss, and P. A. Kaufmann. Ultrafast nuclear myocardial perfusion imaging on

a new gamma camera with semiconductor detector technique: first clinical validation. *European Journal of Nuclear Medicine and Molecular Imaging*, 37(4):773–778, April 2010.

16. J. T. Bushberg, J. A. Seibert, E. M. Leidholdt Jr., and J. M. Boone. *The Essential Physics of Medical Imaging*. LWW, Philadelphia, 3rd edition, December 2011.

17. R. Chandra. *Nuclear Medicine Physics: The Basics*. LWW, Philadelphia, 7th edition, July 2011.

18. S. R. Cherry, J. A. Sorenson, and M. E. Phelps. *Physics in Nuclear Medicine*. Saunders, Philadelphia, 4th edition, April 2012.

19. S. R. Cook, Gelman A., and D. B. Rubin. Validation of software for bayesian models using posterior quantiles. *Journal of Computational and Graphical Statistics*, 15:675692, 2006.

20. T. M. Cover and J. A. Thomas. *Elements of Information Theory*. Wiley-Interscience, Hoboken, NJ, 2nd edition, July 2006.

21. L. Cunha, I. Horvath, S. Ferreira, J. Lemos, P. Costa, D. Vieira, D. S. Veres, K. Szigeti, T. Summavielle, D. Mth, and L. F. Metello. Preclinical imaging: an essential ally in modern biosciences. *Molecular Diagnosis & Therapy*, 18(2):153–173, April 2014.

22. B. De Finetti. *Theory of Probability. A Critical Introductory Treatment. Volume 1*. John Wiley & Sons Ltd, London, New York, 1st edition, January 1974.

23. P. P. Dendy and B. Heaton. *Physics for Diagnostic Radiology*. CRC Press, Boca Raton, 3rd edition, August 2011.

24. R. Diehl, D. H. Hartmann, and N. Prantzos. *Astronomy with Radioactivities*. Springer, Berlin; New York, 2011 edition, October 2010.

25. B. Efron. Bootstrap methods: Another look at the jackknife. *The Annals of Statistics*, 7(1):1–26, January 1979.

26. W. Feller. *An Introduction to Probability Theory and Its Applications, Vol. 1*. Wiley, New York; London, 3rd edition, 1971.

27. W. Feller. *An Introduction to Probability Theory and Its Applications, Vol. 2*. Wiley, New York, 2nd edition, 1971.

28. S. S. Gambhir, D. S. Berman, J. Ziffer, M. Nagler, M. Sandler, J. Patton, B. Hutton, T. Sharir, and S. Ben-Haim. A novel high-sensitivity rapid-acquisition single-photon cardiac imaging camera. *Journal of Nuclear Medicine*, 50(4):635–643, April 2009.

29. A. E. Gelfand and A. F. M. Smith. Sampling-based approaches to calculating marginal densities. *Journal of the American Statistical Association*, 85(410):398–409, June 1990.

30. A Gelman and D. B. Rubin. Inference from iterative simulation using multiple sequences. *Statistical Science*, 7(4):457–472, November 1992.

31. A. Gelman, J. B. Carlin, H. S. Stern, D. B. Dunson, A. Vehtari, and D. Rubin. *Bayesian Data Analysis*. Chapman and Hall/CRC, Boca Raton, 3rd edition, November 2013.

32. S. Geman and D. Geman. Stochastic relaxation, gibbs distributions, and the bayesian restoration of images. *IEEE Transactions on Pattern Analysis and Machine Intelligence*, PAMI-6(6):721–741, November 1984.

33. J. W. Gibbs. *Elementary Principles of Statistical Mechanics*. Ox Bow Pr, Woodbridge, Conn., April 1981.

34. P. Glasserman. *Monte Carlo Methods in Financial Engineering*. Springer, New York, 2003 edition, August 2003.

35. S. N. Goodman. Toward evidence-based medical statistics. 1: The p value fallacy. *Annals of Internal Medicine*, 130(12):995–1004, June 1999.

36. D. Griffiths. *Introduction to Elementary Particles*. Wiley-VCH, Weinheim, 2nd edition, October 2008.

37. G. T. Gullberg, B. W. Reutter, A. Sitek, J. S. Maltz, and T. Budinger. Dynamic single photon emission computed tomography—basic principles and cardiac applications. *Physics in Medicine and Biology*, 55(20):R111, October 2010.

38. T. K. Gupta. *Radiation, Ionization, and Detection in Nuclear Medicine*. Springer, New York, 2013 edition, April 2013.

39. D. I. Hamilton and P. J. Riley. *Diagnostic Nuclear Medicine: A Physics Perspective*. Springer, Berlin; New York, 2004 edition, May 2004.

40. R. W. Hamming. *Numerical Methods for Scientists and Engineers*. Dover Publications, New York, 2nd edition, March 1987.

41. W. K. Hastings. Monte carlo sampling methods using markov chains and their applications. *Biometrika*, 57(1):97–109, April 1970.

42. K. Haven. *Marvels of Science: 50 Fascinating 5-Minute Reads*. Libraries Unlimited, Englewood, Colo, 1st edition, April 1994.

43. M. Hernanz. Novae in gamma-rays. *arXiv:1301.1660 [astro-ph]*, January 2013. arXiv: 1301.1660.

44. W. Huda. *Review of Radiologic Physics*. LWW, Baltimore, MD, 3rd edition, July 2009.

45. M. G. M. Hunink. *Decision Making in Health and Medicine: Integrating Evidence and Values*. Cambridge University Press, Cambridge, 2nd edition, October 2014.

46. J. P. A. Ioannidis. Why most published research findings are false. *PLoS Medicine*, 2(8):e124, August 2005.

47. E.T. Jaynes. On the rationale of maximum-entropy methods. *Proceedings of the IEEE*, 70:939952, 1982.

48. T. Jech. *Set Theory*. Springer, Berlin; New York, 3rd edition, March 2006.

49. H. Jeffreys. An invariant form for the prior probability in estimation problems. *Proceedings of the Royal Society of London. Series A. Mathematical and Physical Sciences*, 186(1007):453–461, September 1946.

50. N. L. Johnson. *Discrete Multivariate Distributions*. Wiley-Interscience, New York, 1st edition, February 1997.

51. N. L. Johnson. *Univariate Discrete Distributions*. Wiley-Interscience,

Hoboken, NJ, 3rd edition, August 2005.

52. N. L. Johnson, S. Kotz, and N. Balakrishnan. *Continuous Univariate Distributions, Vol. 1.* Wiley-Interscience, New York, 1st edition, October 1994.

53. N. L. Johnson, S. Kotz, and N. Balakrishnan. *Continuous Univariate Distributions, Vol. 2.* Wiley-Interscience, New York, May 1995.

54. D. Kahneman and A. Tversky. Prospect theory: An analysis of decision under risk. *Econometrica*, 47(2):263, March 1979.

55. D. Kahneman, P. Slovic, and A. Tversky. *Judgment under Uncertainty: Heuristics and Biases.* Cambridge University Press, Cambridge; New York, 1st edition, April 1982.

56. J. Kataoka, A. Kishimoto, T. Nishiyama, T. Fujita, K. Takeuchi, T. Kato, T. Nakamori, S. Ohsuka, S. Nakamura, M. Hirayanagi, S. Adachi, T. Uchiyama, and K. Yamamoto. Handy compton camera using 3d position-sensitive scintillators coupled with large-area monolithic MPPC arrays. *Nuclear Instruments and Methods in Physics Research Section A: Accelerators, Spectrometers, Detectors and Associated Equipment*, 732:403–407, December 2013.

57. S. Kay. *Fundamentals of Statistical Signal Processing, Volume I: Estimation Theory.* Prentice Hall, Englewood Cliffs, NJ, 1st edition, April 1993.

58. O. Klein and Y. Nishina. ber die streuung von strahlung durch freie elektronen nach der neuen relativistischen quantendynamik von dirac. *Zeitschrift für Physik*, 52(11-12):853–868, November 1929.

59. G. F. Knoll. *Radiation Detection and Measurement.* Wiley, Hoboken, NJ, 4th edition, August 2010.

60. K. S. Krane. *Introductory Nuclear Physics.* Wiley, New York, 3rd edition, October 1987.

61. H. Kunieda, K. Mitsuda, T. Takahashi, R. Kelley, and N. White. Current status of suzaku and its early results. In *Proc. SPIE 6266*, volume 6266, pages 626605–7, Orlando, 2006.

62. K Lange and R Carson. EM reconstruction algorithms for emission and transmission tomography. *Journal of Computer Assisted Tomography*, 8 (2):306–316, April 1984.

63. P. S. Laplace. *Theorie analytique des probabilites;.* Paris, Ve. Courcier, 1812.

64. L. Le Cam. An approximation theorem for the poisson binomial distribution. *Pacific Journal of Mathematics*, 10:11811197, 1960.

65. W. Lee and T. Lee. A compact compton camera using scintillators for the investigation of nuclear materials. In *2009 IEEE Nuclear Science Symposium Conference Record (NSS/MIC)*, pages 641–644, October 2009.

66. E. L. Lehmann and G. Casella. *Theory of Point Estimation.* Springer, 2nd edition, June 1983.

67. E. L. Lehmann and J. P. Romano. *Testing Statistical Hypotheses.*

Springer, New York, 3rd edition, August 2008.

68. S. M. Lesch and D. R. Jeske. Some suggestions for teaching about normal approximations to poisson and binomial distribution functions. *The American Statistician*, 63(3):274–277, 2009.

69. D. V. Lindley. On a measure of the information provided by an experiment. *The Annals of Mathematical Statistics*, 27(4):986–1005, December 1956.

70. S. Matej and R. M. Lewitt. Practical considerations for 3-d image reconstruction using spherically symmetric volume elements. *IEEE Transactions on Medical Imaging*, 15:68–78, February 1996.

71. D. R. McDonald. On the poisson approximation to the multinomial distribution. *Canadian Journal of Statistics*, 8(1):115–118, January 1980.

72. N. Metropolis, A. W. Rosenbluth, M. N. Rosenbluth, A. H. Teller, and E. Teller. Equation of state calculations by fast computing machines. *The Journal of Chemical Physics*, 21(6):1087–1092, June 1953.

73. E. V. Morris, C. J. Enders, K. C. Schmidt, T. C. Bradley, R. F. Muzic Jr., and E. E. Fisher. Kinetic modeling in positron emission tomography. In *Emission Tomography: The Fundamentals of PET and SPECT*, pages 499–540. Academic Press, Amsterdam; Boston, December 2004.

74. H. Müller-Krumbhaar and K. Binder. Dynamic properties of the monte carlo method in statistical mechanics. *Journal of Statistical Physics*, 8 (1):1–24, May 1973.

75. S. B. Mushlin and H. L. Greene II. *Decision Making in Medicine: An Algorithmic Approach, 3e*. Mosby, Philadelphia, PA, 3rd edition, November 2009.

76. M. E. J. Newman and G. T. Barkema. *Monte Carlo Methods in Statistical Physics*. Oxford University Press, Oxford; New York, April 1999.

77. T.E. Nichols, J. Qi, E. Asma, and R.M. Leahy. Spatiotemporal reconstruction of list-mode PET data. *IEEE Transactions on Medical Imaging*, 21(4):396–404, April 2002.

78. N. H. Patel, N. S. Vyas, B. K. Puri, K. S. Nijran, and A. Al-Nahhas. Positron emission tomography in schizophrenia: A new perspective. *Journal of Nuclear Medicine*, 51(4):511–520, April 2010.

79. M. E. Phelps. *PET: Molecular Imaging and Its Biological Applications*. Springer, New York, 2004 edition, May 2004.

80. M.E. Phelps, S-C Huang, E. J. Hoffman, C. Selin, L. Sokoloff, and D.E. Kuhl. Tomographic measurement of local cerebral glucose metabolic rate in humans with (f-18) 2-fluoro-2-deoxy-d-glucose. *Annals of Neurology*, 6:371–388, 1979.

81. J. R. Pierce. *An Introduction to Information Theory: Symbols, Signals and Noise*. Dover Publications, Mineola, NY, 2nd edition, November 1980.

82. W. H. Press, S. A. Teukolsky, W. T. Vetterling, and B. P. Flannery. *Numerical Recipes in C++: The Art of Scientific Computing*. Cambridge

University Press, Cambridge, UK; New York, 2nd edition, February 2002.

83. R.R. Raylman, B. E. Hammer, and N.L. Christensen. Combined MRI-PET scanner: a monte carlo evaluation of the improvements in PET resolution due to the effects of a static homogeneous magnetic field. *IEEE Transactions on Nuclear Science*, 43(4):2406–2412, August 1996.

84. C. P. Robert. *The Bayesian Choice: From Decision-Theoretic Foundations to Computational Implementation*. Springer Verlag, New York, New York, 2nd edition, June 2007.

85. C. P. Robert and G. Casella. *Monte Carlo Statistical Methods*. Springer, New York, 2nd edition, August 2005.

86. M. Rudin. *Molecular Imaging: Basic Principles and Applications in Biomedical Research*. Imperial College Press, London, 2nd edition, September 2013.

87. G. B. Saha. *Basics of PET Imaging: Physics, Chemistry, and Regulations*. Springer, New York, 2nd edition, April 2010.

88. L. J. Savage. *The Foundations of Statistics*. Dover Publications, New York, 2nd revised edition, June 1972.

89. T. Sellke, M. J Bayarri, and J. O. Berger. Calibration of values for testing precise null hypotheses. *The American Statistician*, 55(1):62–71, 2001.

90. SensL_com. Low light silicon photomultipliers (sensl.com), 2014.

91. C.E. Shannon. A mathematical theory of communication. *Bell System Technical Journal*, 27(3):379–423, July 1948.

92. L.A Shepp and Y. Vardi. Maximum likelihood reconstruction for emission tomography. *IEEE Transactions on Medical Imaging*, 1(2):113–122, October 1982.

93. S. Shreve. *Stochastic Calculus for Finance II: Continuous-Time Models*. Springer, New York, NY; Heidelberg, 1st edition, June 2008.

94. A Sitek. Representation of photon limited data in emission tomography using origin ensembles. *Physics in Medicine and Biology*, 53(12):3201–3216, June 2008.

95. A. Sitek. Data analysis in emission tomography using emission-count posteriors. *Physics in Medicine and Biology*, 57(21):6779–6795, November 2012.

96. A. Sitek and S. C. Moore. Evaluation of imaging systems using the posterior variance of emission counts. *IEEE Transactions on Medical Imaging*, 32(10):1829–1839, October 2013.

97. A. Sitek, R. H. Huesman, and G. T. Gullberg. Tomographic reconstruction using an adaptive tetrahedral mesh defined by a point cloud. *IEEE Transactions on Medical Imaging*, 25:1172–1179, September 2006.

98. P. J. Slomka, D. Dey, W. L. Duvall, M. J. Henzlova, D. S. Berman, and G. Germano. Advances in nuclear cardiac instrumentation with a view towards reduced radiation exposure. *Current Cardiology Reports*, 14(2):208–216, April 2012.

99. L. Sokoloff. The (C^{14}) deoxyglucose method: four years later. *Acta Neurologica Scandinavica*, 60:640–649, 1979.

100. J. M. Steele. Le cam's inequality and poisson approximations. *The American Mathematical Monthly*, 101(1):48, January 1994.

101. J. A. C. Sterne and G. D. Smith. Sifting the evidence–what's wrong with significance tests? *BMJ : British Medical Journal*, 322(7280):226–231, January 2001.

102. H. Tajima, T. Kamae, G. Madejski, M. P. Kipac, T. Takahashi, K. Nakazawa, S. Watanabe, T. Mitani, T. Tanaka, Sagamihara Jaxa, Y. Fukazawa, Hiroshima U, J. Kataoka, T. Ikagawa, Tokyo Inst Tech, M. Kokubun, K. Makishima, Tokyo U, Y. Terada, Riken Wako, M. Nomachi, Osaka U, M. Tashiro, and Saitama U. Design and performance of soft gamma-ray detector for next mission. *ECONF C041213:2525,2004*, May 2005.

103. T. Takahashi, S. Takeda, H. Tajima, and S. Watanabe. Visualization of radioactive substances with a si/CdTe compton camera. In *2012 IEEE Nuclear Science Symposium and Medical Imaging Conference (NSS/MIC)*, pages 4199–4204, October 2012.

104. S. Takeda. Experimental study of a si/CdTe semiconductor compton camera for the next generation of gamma-ray astronomy. *Ph.D. Thesis*, page 149, 2009.

105. T. Tanaka, T. Mitani, S. Watanabe, K. Nakazawa, K. Oonuki, G. Sato, T. Takahashi, K. Tamura, H. Tajima, H. Nakamura, M. Nomachi, T. Nakamoto, and Y. Fukazawa. Development of a si/CdTe semiconductor compton telescope. *arXiv:astro-ph/0410058*, pages 229–240, September 2004. arXiv: astro-ph/0410058.

106. A. W. van der Vaart. *Asymptotic Statistics*. Cambridge University Press, Cambridge, June 2000.

107. J-S. Wang and R. H. Swendsen. Cluster monte carlo algorithms. *Physica A: Statistical Mechanics and its Applications*, 167(3):565–579, September 1990.

108. S. Watanabe, T. Tanaka, K. Nakazawa, T. Mitani, Kousuke Oonuki, T. Takahashi, T. Takashima, H. Tajima, Y. Fukazawa, M. Nomachi, S. Kubo, M. Onishi, and Y. Kuroda. A si/CdTe semiconductor compton camera. *IEEE Transactions on Nuclear Science*, 52(5):2045–2051, October 2005.

109. R. Weissleder, B. D. Ross, A. Rehemtulla, and S. S. Gambhir. *Molecular Imaging*. Pmph Usa, Shelton, Conn, 1st edition, June 2010.

110. B. Williams. *Compton Scattering. The Investigation of Electron Momentum Distributions*. McGraw-Hill, 1977.

111. U. Wolff. Collective monte carlo updating for spin systems. *Physical Review Letters*, 62(4):361–364, January 1989.

112. L. Yaroslavtseva. Nonclassical error bounds for asymptotic expansions in the central limit theorem. *Theory of Probability & Its Applications*,

53(2):365–367, January 2009.

113. K. Ye and J. O. Berger. Noninformative priors for inferences in exponential regression models. *Biometrika*, 78(3):645–656, September 1991.

114. K.-P. Ziock, W.W. Craig, L. Fabris, R.C. Lanza, S. Gallagher, B. K P Horn, and N.W. Madden. Large area imaging detector for long-range, passive detection of fissile material. *IEEE Transactions on Nuclear Science*, 51(5):2238–2244, October 2004.

Index

A

Acceptance frequency, 116
Activity, definition of, 89
AE stage, *see* After-experiment stage
After-the-experiment decision making, 42–56
    binary hypothesis testing/detection, 52–56
    Cartesian distance, 44
    example, 42, 45, 47
    interval estimation, 48–50
    maximum a posteriori estimator, 46
    minimum mean square error estimator, 45
    multiple-alternative decisions, 50–52
    point estimation, 43–48
    posterior covariance, 45
    posterior expected loss, 42
    quadratic loss, 44
After-experiment (AE) stage, 3, 41
Alpha particles, 133
Anger logic, 154, 159
Antineutrino, 13
APD, *see* Avalanche photodiode
Astronomy, 176
Autocorrelation
    function, 125
    MC standard error and, 124
Avalanche photodiode (APD), 143, 145, 177

B

Backscatter, 138
Basic statistical concepts, 1–36
    before- and after-the-experiment concepts, 2–6
        after-experiment stage, 3
        before-experiment stage, 3

        examples, 4
        goal of experiment, 3
        known quantities, 2
        observable quantities, 2
        parameter, 3
        quantities of interest, 2
        state of nature, 2
        true values, 3
        unobservable quantities, 2
    BE stage quantities of interests, 35
    data, 1
    definition of probability, 6–10
        countable and uncountable quantities, 8–10
        Dirac delta function, 9
        example, 7
        quantities of interest, 8
    extension to multi–dimensions, 25–29
        chain rule and marginalization, 26–27
        example, 26, 28
        nuisance quantities, 27–29
        vector notation, 25
    frequentist statistics, 1
    histogram of outcomes, 1
    joint and conditional probabilities, 10–13
        conditional distribution, 10
        example, 11, 12
        joint probability distribution, 10
        marginalized probability distribution, 11
    likelihood, 17–19
        example, 17
        likelihood function, 17
        maximum likelihood methods, 18
    loss function, 35

pre-posterior and posterior, 19–25

  Bayesian analysis, 21

  complete objectivity, 20

  designs of experiments, 23–25

  entropy priors, 22

  examples, 22–23

  flat prior, 21

  knowledge gain, 21

  posterior through Bayes theorem, 19–20

  prior selection, 20–22

  reduction of pre-posterior to posterior, 19

  subjectivity, 21

  uninformative prior, 21

pseudo-objectivity of classical statistics, 36

quantities, 1

random variable, 1

statistical model, 13–16

  conditional distributions, 13

  example, 13–14

  motivations, 13

unconditional and conditional independence, 29–34

  conditional independence, 30

  conditional joint, 30

  example, 32, 34

  notation, 31

  scenarios, 31

  unconditional independence, 29

Bayes risk, 57, 58–61

Before-the-experiment decision making, 56–64

  Bayes risk, 57, 58–61

  Kullback–Leibler divergence, 63

  other methods, 62–64

  risk function, 57

  Shannon information entropy, 62

  utility function, 62

Before-experiment (BE) stage, 3, 41

Bernoulli process, 71

Bernoulli trial, 89, 108

BE stage, see Before-experiment stage

Beta particles, 133

Binary hypothesis testing/detection, 52–56

Binomial distribution, 71

Bootstrap, 126–127

Box–Muller transform, 103

Burn-in time, 122

C

Cadmium Zinc Telluride (CZT), 142

CC, see Compton camera

C-C model, see Continuous-to-continuous model

Central limit theorem (CLT), 93, 95–97

Chain thinning, 124

Chemical elements, 131

CLT, see Central limit theorem

Cluster algorithm, 117

Compartmental model, 169–171

Complete data, statistics of, 77–84

Compton camera (CC), 166

Compton scatter, 137

Computed Tomography (CT), 147

Computing, see Statistical computing

Conditional distribution, 10

Conditional independence, 30, 82

Conditional joint, 30

Confidence set, highest posterior density, 49

Continuous-to-continuous (C-C) model, 68

Correlation time, 124

Counting statistics, 67–97

  binomial statistics of nuclear decay, 71–72

    Bernoulli process, 71

    binomial distribution, 71

    known quantities, 72

  complete data, statistics of, 77–84

    complete data, 78

    conditional independence, 82

decay constant, 79
emission counts, 83
example, 78, 81
goal of imaging, 79
models of complete data, 81
notation, 78, 81
simple tomographic system, 78, 84
statistical model of nuclear imaging, 77
system matrix, 78
fundamental statistical law, 69–71
decay constant, 69
known quantities, 70
reminder, 70
statement, 70
multinomial statistics of detection, 72–77
convention, 73
example, 74
notation, 75
tomography, 75
normal distribution approximation, 93–97
approximation of binomial law, 94–95
central limit theorem, 93, 95–97
Gaussian distribution, 94
multi-dimensional central limit theorem, 96
quantities of interest, 96
photon-limited data, exact models of, 71–88
binomial statistics of nuclear decay, 71–72
multinomial statistics of detection, 72–77
Poisson-multinomial distribution of nuclear data, 84–88
statistics of complete data, 77–84

Poisson approximation of nuclear data, 90–93
expectation maximization algorithm, 93
Le Cams theorem, 92
Poisson-multinomial distribution, 91
voxel activity, 93
Poisson statistics of nuclear decay, 88–90
Bernoulli trial, 89
definition of activity, 89
example, 90
Le Cams theorem, 89
Poisson-binomial distribution, 89
Poisson distribution, 88
Poisson limit theorem, 88
quantities of interest, 88
statistical models, introduction to, 67–68
continuous-to-continuous model, 68
discrete-continuous statistical model, 67
discrete-to-discrete model, 67
observable quantities, 67
photon-limited data, 67
state of nature, 67
unobservable quantities, 67
Credible interval, 49
Credible set, 49
CT, *see* Computed Tomography
CZT, *see* Cadmium Zinc Telluride

D

Data quality, evaluation of, 200
Daughter product, 133
D-C statistical model, *see* Discrete-continuous statistical model
D-D model, *see* Discrete-to-discrete model
Dead-time effect, 166
Decay

Bernoulli process, 71
binomial distribution, 71
constant, 69, 79, 133
Decision
    multiple-alternative, 50
    -theoretic procedure, 39
Decision making
    ideal-observer rule of, 55
    medical, 21
Decision theory, elements of, 37–65
    after-the-experiment decision mak-
        ing, 42–56
        binary hypothesis testing/detection,
            52–56
        Cartesian distance, 44
        example, 42, 45, 47
        interval estimation, 48–50
        maximum a posteriori estima-
            tor, 46
        minimum mean square error
            estimator, 45
        multiple-alternative decisions,
            50–52
        point estimation, 43–48
        posterior covariance, 45
        posterior expected loss, 42
        quadratic loss, 44
    Bayes risk, 57, 58–61
        admissibility of Bayes rule, 60
        example, 58, 61
        imaging system, 59
        mean square error, 60
    before-the-experiment decision mak-
        ing, 56–64
        Bayes risk, 57, 58–61
        Kullback–Leibler divergence, 63
        other methods, 62–64
        risk function, 57
        Shannon information entropy,
            62
        utility function, 62
    binary hypothesis testing/detection
        52–56

Bayesian hypothesis testing,
    53
    example, 55
    ideal observer, 53
    likelihood ratio, 54, 55
    loss function, 53
    notation, 53
    unobservable quantity, 53
estimation decision problem, 38
interval estimation, 48–50
    credible interval, 49
    credible set, 49
    definition, 49
    highest posterior density, 49
labels, 38
loss function and expected loss,
    39–42
    decision-theoretic procedure, 39
    example, 39, 40
    expected loss function, 42
    heuristic, 39
    loss function, 39, 41
    optimal decision, 41, 42
    quantities of interest, 41
medical imaging, 38
multiple-alternative decisions, 38,
    50–52
    Bayesian decision making, 51
    composite hypothesis, 51
    detection problem, 50
    membership function, 50
    simple hypothesis testing, 51
observable quantities, 37
robustness analysis, 64–65
    selection of loss function, 64
    selection of prior, 65
    selection of statistical model,
        64
    sensitivity analysis, 64
    unobservable quantities, 37
Depth of interaction (DOI), 164
Detailed balance, 114–115
Detection problem, 50
Dirac delta, 9, 60

Discrete-continuous (D-C) statistical model, 67
Discrete-to-discrete (D-D) model, 67
Discrete Markov chains, 125
DOI, *see* Depth of interaction
Doppler broadening effect, 138
Drug discovery, 174
Dynamic equilibrium, 115
Dynamic imaging, 168–173

E

EC, *see* Emission counts
Effective correlation time, 126
Electron capture, 135
Emission counts (EC), 83
Empty set, 213
Entropy prior, 22, 184
Equation (6.11), 223
Estimation, 38
Estimator
    discrete values, 44
    interval, 49
    MAP, 46
    MMSE, 45
Excited nuclides, 132
Expectation maximization algorithm, 93
Expected loss, 39–42

F

Fisher information matrix, 183
Flat prior, 21, 182
Frequentist statistics, 1
Functional imaging, 147
Fundamental statistical law, 69–71

G

Gamma camera
    Anger logic, 154
    collimator, 155
    imaging with, 153–159
Gamma prior, 182, 198
Gamma prompt photon, 132

Gamma rays, 133
Gaussian distribution, 94
Gibbs sampling, 117
Gravitational forces, 134
Ground state (nucleus), 132

H

Heuristic, 39
Highest posterior density (HPD), 49
Histogram, 1, 104, 105
HPD, *see* Highest posterior density
Hypothesis, definition of, 53
Hypothesis testing
    Bayesian, 53
    composite, 51
    simple, 51

I

IC, *see* Internal conversion
Ideal observer, 53
Ideal-observer (IO) rule of decision making, 55
Image reconstruction, 195–197
Imaging
    count, 150
    event, 150
    functional, 148
    goal of, 79
    molecular, 129
    system, 59
    tomographic, 193
    volume elements, 148
Internal conversion (IC), 134
Interval estimation, 48–50
Intricacy, 152
IO rule of decision making, *see* Ideal-observer rule of decision making
Isomeric transition (IT), 134
Isotopes, 131
IT, *see* Isomeric transition

J

Jackknifing, 126
Jeffreys prior, 183
Joint probability distribution, 10

K

Kinetic energy, 137
KL divergence, *see* Kullback–Leibler divergence
Klein-Nishina formula, 138
Knowledge gain, 21
Known quantities (KQs), 2, 70, 72
KQs, *see* Known quantities
Kronecker, 60
Kullback–Leibler (KL) divergence, 63

L

Labels, 38
Le Cams theorem, 89, 92
LF, *see* Likelihood function
LFI, *see* Location of the first interaction
Likelihood function (LF), 17
Likelihood ratio, 54, 55
Limit cycle, 115
Linear attenuation coefficient, 139
Line of response (LOR), 162
List-mode data, 151
Location of the first interaction (LFI), 162
LOR, *see* Line of response
Loss function, 39
    definition of, 41
    expected, 42
    selection of, 64
Lutetium Yttrium Orthosilicate (LYSO), 143
LYSO, *see* Lutetium Yttrium Orthosilicate

M

Magnetic Resonance Imaging (MRI), 147–148

MAP estimator, *see* Maximum a posteriori estimator
Marginalized probability distribution, 11
Markov chain, 111–127
    autocorrelation, 124
    continuous, 112
    design of, 116–118
    discrete, 125
    equilibrium, 120
    example, 112
    generation of, 186–187
    Markov state, 112
    mixing, 123
    relevance of, 113
    reversible, 115
    state of nature, 112
    thinning, 124
    transition probability, 115
Markov Chain Monte Carlo (MCMC) methods, 181
Markov processes, 113–114
Markov state, 112
Mass number, 131
Maximum a posteriori (MAP) estimator, 46
Maximum likelihood (ML) methods, 18
MCMC methods, *see* Markov Chain Monte Carlo methods
Mean square error (MSE), 60
Membership function, 50
Mesons, 131
Metastable nuclides, 132
Metropolis–Hastings algorithm, 118–120, 181
Minimum mean square error (MMSE) estimator, 45, 197
ML methods, *see* Maximum likelihood methods
MMSE estimator, *see* Minimum mean square error estimator
Molecular imaging, 129, 173
Monte Carlo integration, 107–110

Monte Carlo methods in posterior analysis, 99–127
  continuous distributions, 99–104
    Box–Muller transform, 103
    example, 102
    histogram, 104, 105
    notation, 100
    reminder, 100
    source of error, 101
    Taylor expansion, 101
  design of Markov chain, 116–118
    acceptance frequency, 116
    cluster algorithm, 117
    example, 117
    Gibbs sampling, 117
    selection, 116
    Wolff algorithm, 117
  detailed balance, 114–115
    dynamic equilibrium, 115
    limit cycle, 115
    Markov matrix, 114
    stochastic matrix, 114
    time reversibility, 115
  discrete distributions, 104–107
    assumption, 104
    quantities of interest, 105
  equilibrium, 120–126
    autocorrelation, 124, 125
    burn-in time, 122
    chain thinning, 124
    correlation time, 124
    discrete Markov chains, 125
    effective correlation time, 126
    example, 122
    Markov time, 125
    Markov trace, 124
    poorly mixing chains, 123
    reaching equilibrium, 121
    well-mixing chain, 123
    Z-score, 124
  Markov chains, 111–127
    continuous, 112
    design of Markov chain, 116–118

    detailed balance, 114–115
    equilibrium, 120–126
    example, 112
    Markov processes, 113–114
    Markov state, 112
    Metropolis–Hastings sampler, 118–120
    relevance of, 113
    resampling methods (bootstrap), 126–127
    state of nature, 112
  Markov processes, 113–114
    equilibrium, 114
    ergodicity, 114
    state space, 113
    transition probability, 113
  Metropolis–Hastings sampler, 118–120
    case, 119
    example, 119
  Monte Carlo approximations of distributions, 99–107
    continuous distributions, 99–104
    discrete distributions, 104–107
  Monte Carlo integrations, 107–110
    Bernoulli trial, 108
    law of large numbers, 109
  Monte Carlo summations, 110–111
    example, 110
    quantities of interest, 110
  resampling methods (bootstrap), 126–127
    bootstrap sets, 127
    jackknifing, 126
    resampling method, 127
MRI, *see* Magnetic Resonance Imaging
MSE, *see* Mean square error
Multi-dimensional central limit theorem, 96

Multiple-alternative decisions, 38, 50–52

Multivariate distributions, 211–212

N

Neutrino, 133

NI, *see* Nuclear imaging

Nuclear decay
  alpha particles, 133
  beta particles, 133
  gamma rays, 133

Nuclear imaging (NI), 129–178
  applications, 173–178
    astronomy, 176
    avalanche photodiodes, 177
    clinical applications, 173–174
    drug discovery, 174
    molecular imaging, 173
    pre-clinical development, 174
    radioligands, 174
    Soft Gamma-ray Detector, 177
    standardized uptake value, 174
    tracers, 174, 175
    translational research, 174
  atoms and chemical reactions, 130–131
    atom, 130
    atomic number, 131
    chemical elements, 131
    isotopes, 131
    mass number, 131
  coincidence detection, 161–162
    coincidence, 161
    energy window, 162
    random coincidences, 161–162
    scatter coincidence errors, 161
    single detection, 161
  compartmental model, 169–171
    example, 169
    input function, 170
    interpretation, 170
  Compton camera, 166
  Compton imaging, 166–168

dynamic imaging and kinetic modeling, 168–173
  compartmental model, 169–171
  dynamic measurements, 171–173
  functional methods, 168
  region of interests, 169
gamma camera, imaging with, 153–159
  Anger logic, 154
  collimation, 155
  gamma camera, 153–157
  parallel hole collimator, 156
  SPECT, 157–159
inelastic scattering, 137–138
  backscatter, 138
  Doppler broadening effect, 138
  kinetic energy, 137
  Klein-Nishina formula, 138
  recoiled electron, 137
interaction of radiation with matter, 136–141
  inelastic scattering, 137–138
  photoelectric effect, 138
  photon attenuation, 138–141
molecular imaging, 129
nuclear imaging, 147–168
  activity concentration, 149
  Becquerel, 149
  Compton imaging, 166–168
  Computed Tomography, 147
  example, 149
  functional imaging, 147
  imaging with gamma camera, 153–159
  Magnetic Resonance Imaging, 147–148
  photon-limited data, 150–151
  positron emission tomography, 159–166
  quantitation of PET, 165–166
  radiodensity, 148
  radium isotope, 149

region of response, 152–153
Ultrasound, 148
volume elements, 148
nuclear radiation, 130–141
  atoms and chemical reactions, 130–131
  basics of nuclear physics, 130–136
  nucleus and nuclear reactions, 131–133
  types of nuclear decay, 133–136
nucleus and nuclear reactions, 131–133
  binding energy, 132
  daughter product, 133
  decay constant, 133
  decay scheme, 132
  excited nuclides, 132
  gamma prompt photon, 132
  ground state, 132
  mesons, 131
  metastable nuclides, 132
  nuclear strong interaction, 131
  nucleons, 131
  nuclide, 132
  radioactive decay, 132, 133
  residual strong force, 131
  stable nucleus, 132
photon attenuation, 138–141
  example, 140
  Thompson scattering, 139
photon-limited data, 150–151
  count, 150
  event, 150
  list-mode data, 151
positron emission tomography, 129, 159–166
  coincidence detection, 161–162
  PET nuclear imaging scanner, 159–161
  quantitation of PET, 165–166
  ROR for PET and TOF-PET, 162–164
quantitation of PET, 165–166

attenuation, 165
Compton scatter and randoms, 166
dead time, 166
normalization, 165
radiation detection in nuclear imaging, 141–147
  avalanche photodiode, 143, 145
  Cadmium Zinc Telluride, 142
  count rate, 142
  detector efficiency, 142
  Geiger–Mller detector, 146
  Lutetium Yttrium Orthosilicate, 143
  photomultiplier tubes, 143, 144–145
  photon counting detectors, 141
  scintillation detectors, 143–147
  semiconductor detectors, 142
  solid-state photomultipliers, 145–147
radionuclides, 129
radiopharmaceutical, 129
ROR for PET and TOF-PET, 162–164
  depth of interaction, 164
  line of response, 162
  location of the first interaction, 162
  parallax error, 164
single photon emission computed tomography, 129
statistical model of, 77
types of nuclear decay, 133–136
  alpha particles, 133
  antineutrino, 13
  beta particles, 133
  electron capture, 135
  gamma rays, 133
  gravitational forces, 134
  internal conversion, 134
  isomeric transition, 134
  neutrino, 133
  positron, 135

weak forces, 134
Nucleons, 131
Nucleus, 131–133
    binding energy, 132
    daughter product, 133
    decay constant, 133
    decay scheme, 132
    electron capture, 135
    excited nuclides, 132
    gamma prompt photon, 132
    ground state, 132
    internal conversion, 134
    isomeric transition, 134
    mesons, 131
    metastable nuclides, 132
    nuclear strong interaction, 131
    nucleons, 131
    nuclide, 132
    positron, 135
    radioactive decay, 132, 133
    radionuclide, 133
    residual strong force, 131
    stable, 132
    strong interaction, 131
    weak nuclear force, 134
Nuclide, definition of, 132
Nuisance QoIs, 27

O

Observable quantities (OQs), 2
    counting statistics, 67
    decision theory, 37
    loss function and expected loss,
        41
    statistical model, 15
OE algorithm, see Origin Ensemble
        algorithm
OE algorithm (for STS), C++ code
        of, 225–230
Optimal decision, 41, 42
OQs, see Observable quantities
Origin Ensemble (OE) algorithm,
        179–180

P

Parallax error, 164
Parallel hole collimator, 156
PET, see Positron emission tomogra-
        phy
Photoelectric effect, 138
Photomultiplier tubes (PMTs), 143,
        144–145
Photon
    attenuation, 138–141, 165
        Compton scatter, 137
        linear attenuation coefficient,
            139
        photoelectric effect, 138
        scatter, 136
        Thompson scattering, 139
    counting detectors, 141
    energy, 142
    gamma prompt, 132
    -limited data (PLD), 67, 71–88,
        150–151
    secondary, 142
PLD, see Photon-limited data
PMTs, see Photomultiplier tubes
Poisson-binomial distribution, 89
Poisson distribution, 88
Poisson limit theorem, 88
Poisson-multinomial distribution, 85,
        91
Poorly mixing chains, 123
Positron annihilation, 135
Positron emission tomography (PET),
        129, 159–166
block detectors, 159
coincidence, 161
    crystals, 159
depth of interaction, 164
line of response, 162
    non-collinearity, 159
    positron range, 159
quantitation, 165–166
randoms, 166
scatter, 163, 166
single, 163

Posterior analysis, *see* Monte Carlo methods in posterior analysis

Posterior covariance, 45

Posterior expected loss, 42

Probability, definition of, 6–10

Probability distributions, 209–212
  multivariate distributions, 211–212
    multinomial distribution, 211
    multivariate normal distribution, 211
    Poisson-multinomial distribution, 211–212
  univariate distributions, 209–211
    binomial distribution, 209
    gamma distribution, 209
    negative binomial distribution, 209
    Poisson-binomial distribution, 210
    Poisson distribution, 210
    uniform distribution, 210
    univariate normal distribution, 210–211

Q

QoI, *see* Quantities of interest

Quadratic loss, 44

Quantities of interest (QoI), 2
  BE stage, 35
  central limit theorem, 96
  discrete distributions, 105
  loss function and expected loss, 41
  Monte Carlo summations, 110
  multinomial statistics of detection, 75
  multiple-alternative decisions, 51
  nuisance, 27
  Poisson-multinomial distribution of nuclear data, 85
  Poisson statistics of nuclear decay, 88

statistical model, 14

R

Radiation detector
  count rate, 142
  efficiency, 142
  photon counting, 141
  scintillation detectors, 143–147

Radioactive decay, 132, 133

Radiodensity, 148

Radioligands, 174

Radionuclides, 129

Radiopharmaceutical, 129

Random variable, 1

Recoiled electron, 137

Region of interest (ROI), 169, 203

Region of response (ROR), 152–153, 179

Resampling methods (bootstrap), 126–127

Residual strong force, 131

Risk function, 57

Robustness analysis, 64–65

ROI, *see* Region of interest

ROR, *see* Region of response

S

Sampling distribution ratios, derivations of, 221

Scintillation detectors, 143–147

Sensitivity analysis, 64–65

Set theory, 213–215
  cardinality, 213
  definition of set, 213
  disjoint sets, 214
  elements, 213
  empty set, 213
  intersection, 214
  partition, 214
  proper subset, 214
  subset, 214
  superset, 214
  union, 214

Venn diagrams, 214
SGD, *see* Soft Gamma-ray Detector
Shannon information entropy, 62
Simple tomographic system (STS), 78, 84, 194–195
Single photon emission computed tomography (SPECT), 129, 157–159
 assumption, 158
 projection, 157
 tomographic information, 157
Single-voxel imaging, multinomial distribution of, 217–219
Soft Gamma-ray Detector (SGD), 177
Solid-state photomultipliers, 145–147
SoN, *see* State of nature
SPECT, *see* Single photon emission computed tomography
Standardized uptake value (SUV), 174
State of nature (SoN), 2
 counting statistics, 67
 Markov chains, 112
 partial knowledge of, 10
Statistical computing, 179–207
 computationally efficient priors, 182–185
  entropy prior, 184
  Fisher information matrix, 183
  flat prior, 182
  gamma prior, 182
  Jeffreys prior, 183
  notation, 182
 examples of statistical computing, 193–207
  Bayes factors, 198–200
  Bayes risk, 205–207
  bootstrapping, 204
  detection (Bayesian decision making), 203–205
  evaluation of data quality, 200–203
  experiment, 195
  gamma prior sensitivity, 198

image reconstruction, 195–197
implementation of gamma prior, 198
loss function, 203
marginalized posteriors, 199
minimum mean square error estimate, 197
OE algorithm, 196
region of interest, 203
simple tomographic system, 194–195
origin ensemble algorithms, 190–193
 acceptance ratio, 190
 expectation, 191
 failure, 192
 negative binomial distribution, 192
 for sampling, 190, 192
Poisson-multinomial statistics, 179–193
 computationally efficient priors, 182–185
 generation of Markov chain, 186–187
 Metropolis–Hastings algorithm, 187–190
 origin Ensemble algorithm, 179–180, 190–193
 sampling the posterior, 180–181
region of response, 179
sampling the posterior, 180–181
 Markov Chain Monte Carlo methods, 181
 Metropolis–Hastings algorithm, 181
 notation, 181
 unobservable quantity, 180
Stochastic matrix, 114
STS, *see* Simple tomographic system
Subset, definition of, 214
SUV, *see* Standardized uptake value
System matrix, 78

T

Thompson scattering, 139
Time reversibility, 115
Tomography, 75
Tracers, 175
Transition probability, 113
Translational research, 174

U

Ultrasound (US), 148
Unconditional independence, 29
Uninformative prior, 21
Univariate distributions, 209–211
Unobservable quantities (UQs), 2
    before-the-experiment decision making, 57
    binary hypothesis testing/detection, 53
    counting statistics, 67
    decision theory, 37
    designs of experiments, 23–25
    loss function and expected loss, 41
    point estimation, 43
    Poisson-multinomial distribution of nuclear data, 84
    prior selection, 21
    sampling the posterior, 180
    statistical model, 15
UQs, *see* Unobservable quantities
US, *see* Ultrasound
Utility function, 62

V

Venn diagrams, 214
Volume elements, 148
Voxel activity, 93

W

Weak forces, 134
Well-mixing chain, 123
Wolff algorithm, 117

Z

Z-score, 124